Plumbing for Beginners

A Comprehensive Guide to Home Plumbing

(Keep Your Home's Plumbing System Works Safely Information Sinks Bathtubs)

Jose Kelch

Published By **John Kembery**

Jose Kelch

All Rights Reserved

Plumbing for Beginners: A Comprehensive Guide to Home Plumbing (Keep Your Home's Plumbing System Works Safely Information Sinks Bathtubs)

ISBN 978-1-7753142-1-9

No part of this guidebook shall be reproduced in any form without permission in writing from the publisher except in the case of brief quotations embodied in critical articles or reviews.

Legal & Disclaimer

The information contained in this book is not designed to replace or take the place of any form of medicine or professional medical advice. The information in this book has been provided for educational & entertainment purposes only.

The information contained in this book has been compiled from sources deemed reliable, and it is accurate to the best of the Author's knowledge; however, the Author cannot guarantee its accuracy and validity and cannot be held liable for any errors or omissions. Changes are periodically made to this book. You must consult your doctor or get professional medical advice before using any of the suggested remedies, techniques, or information in this book.

Upon using the information contained in this book, you agree to hold harmless the Author from and against any damages, costs, and expenses, including any legal fees potentially resulting from the application of any of the information provided by this guide. This disclaimer applies to any damages or injury caused by the use and application, whether directly or indirectly, of any advice or information presented, whether for breach of contract, tort, negligence, personal injury, criminal intent, or under any other cause of action.

You agree to accept all risks of using the information presented inside this book. You need to consult a professional medical practitioner in order to ensure you are both able and healthy enough to participate in this program.

Table Of Contents

Chapter 1: Importance Of Timely Plumbing Repairs 1

Chapter 2: Plumbing Mole Grips 9

Chapter 3: How Does A Flush Toilet Work? 20

Chapter 4: Clean The Pump And Valve ... 31

Chapter 5: Fixing A Clogged Shower Head 40

Chapter 6: Fixing A Leaky Kitchen Tap Or Sprayer 53

Chapter 7: Repairing A Broken Underground Drainage Pipe 67

Chapter 8: Why Plumbing Maintenance Is Important 81

Chapter 9: Understanding The Water System In Your Home 93

Chapter 10: Diy Kitchen Repairs 101

Chapter 11: Diy Kitchen Installations ... 113

Chapter 12: Diy Toilet Repairs 124

Chapter 13: Diy Bathroom Repairs....... 150

Chapter 14: Diy Plumbing Repairs Outside Your Home.. 166

Chapter 1: Importance Of Timely Plumbing Repairs

Many owners won't understand the significance of a properly on foot plumbing device till troubles rise up. While a few plumbing troubles cease cease result from forgetting to keep your household plumbing, others stand up due to the natural growing older of plumbing additives. Regardless of the reason of the plumbing troubles, getting them fixed proper now's important because of the fact:

Prevents Unwanted Disruptions

When handling plumbing problems including clogged drains and flooding toilets, you may need to expose off the water deliver to your home until after resolving the hassle. This can also moreover save you you from using the rest room or wearing out every day domestic duties like doing laundry or washing the dishes.

Timely plumbing renovation assist maintain everyday family utilities and sports strolling easily through stopping undesirable disruptions.

Protects Your Home from Further Damage

Your domestic is honestly certainly one of your most precious belongings. As such, you ought to do the whole thing to keep its fee. If left unattended, plumbing troubles along with leaky taps and busted water pipes can bring about essential water damage in your partitions and floors. This can compromise structural stability, ensuing in pricey building maintenance.

By keeping water from getting access to areas of the house that need to stay dry, early plumbing protection are an effective method to shield your home.

Protects Occupants from Mold Related Health Problems

Mold infestation is the maximum big health risk of a leaky plumbing tool. Asthmatics and people with hypersensitive reactions go through the most whilst there's mould in their houses.

Mold spores are almost anywhere, however they simplest make bigger on damp surfaces. Plumbing leaks can dampen your own home's interior surfaces, presenting the suitable habitat for mould to thrive. This locations your family's fitness in jeopardy.

It's important to restore plumbing leaks as brief as viable to maintain your property dry and reduce the danger of mildew infestation.

It Saves You Money

Apart from the coins you keep with the beneficial useful resource of doing plumbing safety yourself in choice to hiring a person, being timely together with your protection will prevent coins. This is because most plumbing problems get worse if left unattended. For instance, a blocked drainpipe from the bathing device must cause the device itself to malfunction. Therefore, to avoid useless damages and charges, it's miles continuously practical to do your renovation proper away.

By gaining knowledge of a way to do plumbing preservation your self, you may restoration troubles proper now once they rise up in place of searching in advance to a plumber to come back on your rescue.

Let's now skip to the following financial disaster and notice what tools you want to have in your property to recovery not unusual plumbing issues.

Must-Have Plumbing Tools

While solving plumbing troubles spherical your property may also sound easy and fun, you could however need to cope with the preservation with loads of warning. Otherwise, you may come to be with a dysfunctional plumbing device.

For that purpose, residence proprietors want to have a hard and fast of plumbing device at hand to make sure they could do each restore properly. Remember, you cannot have the outcomes of a pro if you do now not have the proper device for the task.

Let's take a look at the numerous gadget residence proprietors need for DIY plumbing upkeep.

Plastic Tube Cutter

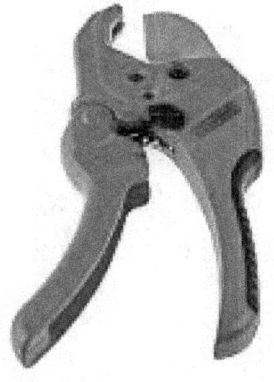

This device allows plumbers lessen thru plastic tubes. The plastic tube cutter comes in severa paperwork, but they continually embody a pointy blade that slices thru plastic tubing. If you want to do plumbing repairs that comprise decreasing plastic water pipes, you can want this tool.

After using strain to a tube cutter, the pointy blade cuts via the plastic tubing. All you need to do is squeeze the two handles inside the course of each other. Alternatively, you can roll the cutter's blade round your plastic pipe circularly, reducing farther via the plastic with each flip.

Metal Pipe Cutter

Normally, maximum tube cutters reduce thru plastic pipes. However, a plastic tube cutter will no longer do while you want to cut metal pipes. Instead, you want to apply a metal pipe cutter.

A pipe cutter operates through slowly lowering thru the metal pipe and severing it with a pointy reducing wheel. The reducing disc slices via the pipe at the same time as the pipe cutter rotates in a circle spherical it. The cutter penetrates deeper into the pipe with every rotation till it cuts it completely.

Hack Saw

A hand-powered hacksaw facilitates plumbers reduce metal pipes, bars, flanges, and similar fittings. This device can also cut plastic pipes. The period of the determined's frame is extendable with a tightening nut or knob, which applies force at the blade and holds it in place. Moreover, you could set the blade to cut at the pushing and pulling stroke, with the frenzy stroke being the most popular.

Hole Saw Kit

A hollow-located is a plumbing device used to reduce exactly round holes in unique plumbing fittings. It can reduce holes with a ways larger diameters than regular drill bits and is a lot greater powerful than extraordinary drilling device as it actually cuts out the outer line of the hollow.

The hollow noticed wants to connect to a power drill that makes it spin at a short speed. Consequently, the enamel of the noticed lessen into the chosen turning into as it revolves, putting off a round chunk (waste) due to the fact it is going.

Tube and Pipe Benders

A pipe or tube bender allows bend various plumbing pipes. For instance, it's far now not possible to apply a at once pipe whilst a pipe has to skip thru severa obstructions which includes stairs. Thus, the pipe needs bending at precise angles to bypass up or down the steps. Luckily with a pipe bender, this undertaking need to be easy to accomplish.

One of the exceptional benefits of bending pipes is its functionality to do away with the want to apply pipe fittings to attach pipe pieces. Remember, it's far better to apply a bent, non-prevent pipe in region of reducing the pipe and connecting it the use of fittings due to the fact you get rid of any probabilities of water leakage.

Chapter 2: Plumbing Mole Grips

A mole grip —now and again known as a locking plier— is a hand-held tool with adjustable jaws that lock round an item to maintain it firmly in place. Mole grips are distinguished from everyday pliers by manner of the use of their locking mechanism, which calls for the customer to keep them closed. Besides firmly conserving pipes and certainly one of a kind furnishings, mole grips also can help loosen numerous screws, bolts, and nuts.

Plumbing Pliers

These pliers have big handles and extendable jaws that lock in location at the contact of a button, making them the most typically used tool in any plumber's toolbox. They're one of the maximum adaptable plumber's gear, with a massive range of beginning sizes and the ability to capture nearly any form. They can tighten or loosen nearly whatever.

It's proper to have plumbing pliers: one for stabilizing the pipe and the other for loosening and tightening. The 10" duration need to meet most of your plumbing desires, but it's correct to have quite some sizes.

Plumber's LED Flash Light

Some factors of your plumbing tool can be in hidden and dark locations, together with inside the basement, at the ceiling, and lots of others. Such locations rarely have exposure to herbal lights. For this cause, it is vital to have a flashlight to offer lighting fixtures whilst you're doing preservation in such locations. While any everyday flashlight can serve the cause, it's wonderful to have particular flashlights made for plumbers.

The plumber's flashlight should be small and compact, turning into into even the tiniest places. However, it ought to despite the fact that provide extreme illumination, just like large, heavier flashlights.

Most of these flashlights embody bendy straps that help you function them everywhere and although rotate them at a 360-diploma attitude. More importantly, they may be water-proof against guard them from any splashes.

Pipe Threading Kit

Pipe threaders are one of the maximum vital machine. They help cut threads at the ends of tubes and pipes to create each lady and male attachments with precision. They constantly are available in handy even as growing a turning into a member of or greater pipes.

If you require a pipe threader for your plumbing repair, you should select one in order to complete the mission correctly. The device is available in severa sizes, starting from heavy-obligation, enterprise-grade machines capable of threading sizeable pipes or greater to small, hand-cranked threaders able to threading thinner pipes.

Most pipe threaders are low-price and designed to offer pinnacle-rated output, even for the hardest pipes.

Pipe Wrench

A pipe wrench is a device made from aluminum or metallic used by plumbers to hold, loosen, or tighten spherical pipes. This tool has a flat address, a pinnacle hook jaw, and a decrease heel jaw element. The lower jaw is movable up or right down to healthy the diameter of the pipe. The jaw section comes with microscopic teeth or holes for holding round pipes whilst have become.

To assist benefit the right grip amount and save you horrific the pipe, there must be a 0.Five-inch place among the rear of the hook jaw and the floor. Leaving this place furthermore prevents you from hurting yourself or detrimental the wrench.

Sink/Basin Wrench

If you've ever attempted tightening a unfastened nut on a everyday spanner, you understand that it's miles an not possible assignment. Luckily, there can be the choice of the usage of a sink wrench, furthermore known as a basin wrench. This is a ought to-have tool for every professional and novice plumbers.

It has an extended manipulate and a revolving, self-adjusting, clutching head. Its essential use is fastening or eliminating faucet tailpiece mounting bolts. The tool's layout permits it to art work in smaller areas in which exceptional gear can't.

Plungers

The plunger enables without trouble unclog blocked pipes and drains. This tool is straightforward-to-use, even in case you don't understand some thing approximately plumbing. Normally, a plunger has a pole and a half of of of-cup made from rubber related to as a minimum one stop. A plunger is a device every assets owner should have because it facilitates remedy minor blockage issues in mins.

Most proprietors use plungers to smooth blocked sinks. However, did you understand that you could furthermore use this device to unblock clogged bathrooms and shower drains? Yes, that's right, and that is why it's far a want to-have.

Hand Auger

A hand auger is a drain-clearing plumbing device operated by hand. It comes with a 25-foot flexible steel twine used to unclog tubs, sinks, showers, and drain pipes. If the purpose of the blockage is a strong however moderate item, which includes tree roots or glass fiber, the tool will break up it open and allow the waft to keep.

In addition, the auger can also snag a smaller, lighter clog, allowing you to pull it out. Some people use this device to unblock clogged lavatories even though it is not useful because of the fact the auger may also go away scratches on the rest room bowl.

Drain Camera

A sewage inspection camera lets in decide whether or not sewer lines are obstructed, collapsed, or fractured. A drain virtual digicam is the very extraordinary way to make sure that no tree roots have located their way into your own home's sewage pipes. After all, sewer traces are pungent and disgusting to address every different manner.

The video digital camera captures the sewer's situation, exposing any fractures, tree roots, broken traces, obstructions, and specific problems. Therefore, in case you suspect a blockage for your sewage pipes, use a virtual digital camera to understand precisely in which the blockage is.

Plumbing Gloves

DIY plumbing maintenance may furthermore sound smooth, but you have to protect your self with gloves regardless. Working with pipes made from likely historical metals, together with copper exposes you to the danger of cuts from lacerated edges. In addition, managing heat pipes and the usage of hand tools may be hard in your palms, necessitating hard plumber paintings gloves designed absolutely for plumbing repair duties.

Although a few plumbers keep away from wearing gloves due to the fact they're capable of impair accuracy, you need them to guard your palms from scrapes and burns. Disposable gloves can be beneficial for a few repair duties, however cloth gloves generally deliver extra protection.

Safety Googles

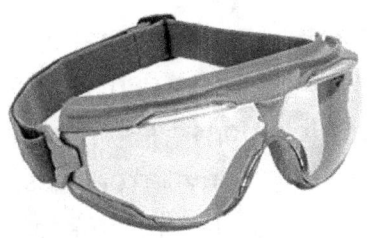

Even as you do simple plumbing repairs round your own home, it is always useful to wear protection goggles to guard your eyes from debris, dust, smoke, and distinct risky chemical substances. To located it some unique way, we placed on safety glasses as a crucial initial line of protection for our eyes which is probably sensitive body organs.

Thus, invest in an exquisite pair of protecting eyeglasses that don't fog. Safety goggles are a have to-have, whether or not or now not you're genuinely reducing pipes or working in a dusty attic.

Tape Measure

A smooth measuring tape is the very last plumbing device wanted by way of manner of the use of each plumber. When dealing with pipes, precise measurements are often required to ensure right protection in constrained locations. A tape degree will assist save you costly errors resulting from defective period or distance measurements.

Remember, there may be no prevent to the listing of plumbing machine to be had these days. However, this listing is complete, especially regarding the basics and the most used plumbing gear. More importantly, you do not have to buy maximum of those equipment right now. You can continuously buy the tools you need in your modern restore responsibilities and slowly broaden your tool series.

Now that we apprehend numerous plumbing system and their makes use of, allow's go with the flow right now to the subsequent financial wreck and take a look at the severa upkeep you can do the usage of the ones system.

Toilet Repairs

Some of the most commonplace plumbing issues in a domestic revolve round the bathroom. This bankruptcy has the entirety you want to recognize about relaxation room plumbing protection. Before we get into the protection, permit us to first understand how a bathroom works.

Chapter 3: How Does A Flush Toilet Work?

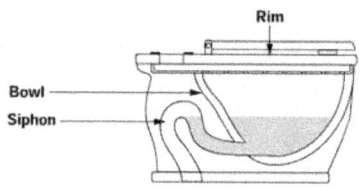

The toilet bowl and the tank are the 2 number one components of a gravity-flush bathroom. Let's start with the bowl, then bypass at once to the tank.

Thanks to its clean however extraordinary layout, the bowl is the most enormous part of your relaxation room as it allows smooth waste disposal through a siphon. When glancing at the factor of the bowl, you'll observe a u-fashioned detail that hyperlinks to the bowl and extends into the floor. This is the siphon, the aspect of the bathroom bowl liable for flushing stuff down into the sewer.

If you took physics in university, you may likely have decided out about siphons. It is a time period used to refer to any conduit that consists of liquid upwards from a massive reservoir after which downwards through growing a vacuum. Gravity handles the relaxation after a super amount of liquid pumps into the reservoir, propelling the water up the U form and thru the pipe. Because water molecules preserve close together, the siphon creates a vacuum that pushes the relaxation of the contents down that pipe as fast as water starts filling that U-shaped region.

Because of the siphon, even in case you put off the tank out of your relaxation room and exceptional have the bowl phase, you can although have a very useful rest room. Notably, it wouldn't make an entire lot distinction in case you slowly poured a cup of water into the bowl. However, in case you took a -gallon pail of water and emptied it into the rest room bowl, gravity might also want to take over and flush the water away.

The U-form on a lavatory bowl furthermore acts as a seal, stopping gasses from the relaxation room bend and sewer from entering your home thru the bathroom. The flushing stops on the equal time as air goes into the siphon, and the rest room bowl fills decrease returned up with water from the tank.

What is the characteristic of the tank?

The tank competencies in addition to a -gallon bucket being poured into the bowl, excellent greater exactly, and it accurately refills itself. The flushing method starts with a push of the tank's manipulate. Pressing the address raises a lever joined to a series.

This chain connects to a rubber flap at the tank's bottom even as the rubber flapper attaches to the tank's seat. The aspect is; the flapper is accountable for developing a seal some of the relaxation room bowl and tank water. When you press the cope with, it attracts the rod up, breaching the seal and permitting the water in the tank to spill into the bowl, developing a flushing impact.

After every flush, the supply valve inside the tank allows water in as plenty because the fill valve, which starts offevolved filling the tank with water all over again. The flapper then returns to its actual function, sealing the tank, due to this preventing any further water from entering into the bowl. The fill valve permits water to enter the tank until the glide reaches the favored diploma, at which factor it closes the fill valve.

To simplify all this, the rest room works in three steps:

The siphon triggers even as the tank empties gallons of water into the bathroom bowl. It then pushes waste and water down into the drain and out to the sewer through gravity. The tank then refills with glowing water, organized for the subsequent flush.

Now which you recognize how a rest room works, allow's check how you may recovery a number of the maximum common bathroom plumbing problems.

Fixing a Toilet That Does Not Flush Fully

One of the maximum ordinary relaxation room troubles is desiring to push the flush lever completely right down to flush. The most possibly motive of the trouble is an excessive amount of anxiety in the lifting chain connecting the flush lever to the flapper.

When the chain is clearly too slack, it can't increase the flap immoderate sufficient to permit the desired quantity of water to run down thru the flush valve. As a cease end result, the flapper closes upfront, interrupting the flush. To repair this trouble, follow the steps beneath:

Examine the chain period to appearance if it goals adjusting: Look within the rest room tank for the chain. Adjust the length of the chain till only 1/2 of inch of slack stays.

Try flushing the toilet another time: At this detail, your lavatory must truely flush pretty surely. However, if it nonetheless doesn't paintings, re-alter the chain.

See if the chain goals trimming: It's viable that the chain has stretched and come to be too extended and is now dangling. This method you need to trim and shorten it the use of decreasing pliers. Just ensure it does no longer hang down too a protracted manner to intrude with every one-of-a-kind rest room detail. By so doing, your rest room should flush nicely.

Stop Your Toilet from Running after a Flush

After a flush, water need to forestall flowing into the bowl. However, in case your relaxation room is faulty, it'll keep strolling, and water will hold to circulate the bowl indefinitely. This hassle can be expensive because it has the ability to waste many gallons of water if no longer addressed.

When this kind of problem occurs, it specially method that the flapper has a hassle due to the reality water will preserve flowing down into the bathroom bowl if the flapper does no longer properly reseal towards the flush valve. After all, the water stage never receives to the element required to lessen off the water deliver valve. Fortunately, you may restore this hassle through manner of following the ones simple steps:

Check to peer if the period of the raise chain goals decreasing: Make positive the carry chain doesn't get caught many of the flapper and the flushing valve. If that is the case, water will circulate the relaxation room bowl after flushing. Reduce the duration of the chain in order that it does no longer get pinched below the flapper.

Confirm if the flapper is properly aligned

Make certain the flapper is perfectly located regular with the flush valve starting, so it closes effectively. You might be able to stop the leakage through way of using virtually repositioning the flapper.

See if the flapper wants to changing: The flapper's rubber can in no way be capable of prevent the water flow into the rest room tank if it's far vintage and damaged. In the shape of case, you haven't any preference however to exchange the flapper.

Tightening a Loose Toilet Seat

The not unusual sitting, final, and setting up of the bathroom seat frequently loosen the bolts that regular your lavatory seat, ensuing in a loose relaxation room seat. Fixing this one is straightforward. The best device you need for the hobby is a screwdriver. In rare cases, you would likely need more than one pliers and an adjustable wrench.

Remember, if the seat is in horrible scenario, now could be an exceptional possibility to shop for and installation a contemporary one. However, if your lavatory seat within reason new but feels loose, have a examine the stairs under to restore the hassle.

Locate the Bolts: Some bolts are visible, but most are hidden through the use of a plastic flap that clamps near. To get proper of get entry to to the bolts that join the seat to the pinnacle of your lavatory bowl, peel those plastic covers open with a screwdriver.

Tighten every bolt: If the nuts have grooved heads, fasten them with a screwdriver by way of way of twisting them clockwise until they may be business organization. Tighten the nuts on every side gently to make certain your relaxation room seat is on an excellent level. If the bolt absolutely spins earlier than tightening, hold close the nut screwed onto the bolt with pliers from underneath the relaxation room whilst stiffening the bolt with a screwdriver from above. Most rest room seats use metallic screws, however if yours has plastic screws, be cautious now not to interrupt them or peel the threads.

Tighten cussed screws from the lowest up: You can modify the attachment nut from the decrease side if important. Turn the bathroom seat nuts round from under the bowl until they'll be corporation. Tightening those bolts is usually quality with a ratchet wrench that has a deep socket, however an adjustable wrench can also paintings.

If now not anything works, update the bolts: Sometimes, your modern bolts can also wreck or refuse to tighten no matter how tough you try. Luckily, you can find out new bolts at a hardware keep or home renovation keep. Frozen bolts may additionally require a sawing blade to eliminate because the blade is so narrow and can healthful beneath the pinnacle of the bolt. That way, it is able to reduce the bolt without scratching the ceramic at the bowl. After the utilization of recent bolts, your loose rest room seat want to experience business enterprise, much like the number one day after set up.

Fixing a Toilet that Fills Up Slowly

Make high-quality the water deliver close-off is definitely open

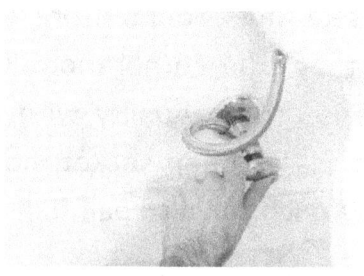

A lavatory tank normally replenishes in 3 mins— depending for your water pressure. If your bathroom tank is taking a long time to fill or isn't filling, start with the toilet supply close to-off. For finest water glide, ensure your water close-off is virtually open. Your slow-filling troubles may be because of it no longer being honestly open.

Chapter 4: Clean The Pump And Valve

If you've examined the water supply and it's although no longer taking walks, smooth the pump and valve due to the truth debris should have accumulated through the years, and an in depth cleansing can be crucial to treatment the problem. If there are mineral buildups at the outdoor of the fill valve, clean it with vinegar and a toothbrush. Wash the valves with heat water and cleaning soap as soon as the accumulation has come off. After that, deliver it an first-rate rinse.

Plunging a Clogged Toilet

To comprehend in case your rest room is clogged, have a examine if it's far draining waste very slowly. You will phrase that the flush water in element fills the relaxation room bowl but does not rush out to wipe up the waste. The water degree remains immoderate for a while in advance than draining to its normal diploma. Therefore, you can no longer phrase that the relaxation room is blocked till you flush it.

Here is a easy device to repair this hassle:

Invest within the right plunger: Anyone can unclog approximately 90% of clogged bathrooms with the proper relaxation room plunger. Make excessive high-quality you purchase a plunger with a rubber, bell-fashioned forestall with an extension flange. This is because it fits bathrooms better, permitting you to offer the plunge greater "oomph."

Start Plunging

The only method to clean "natural" blockages is to plunge—due to the fact a plunge may additionally moreover bring about splashes, area towels round the rest room's base and relocate superb objects to a safe, dry spot earlier than plunging. Leaving a entire rest room to sit down for 20 or half of-hour will often permit plenty of the water to drain to a extra viable degree. The bowl wishes to have enough water to cover the plunger fully.

To make a better closure with the get right of entry to at the bottom of the bowl, fold out the tail from in the plunger. Pump the plunger a 1/2 of-dozen instances, then relaxation for a minute in advance than repeating. Do this for 10 to fifteen rounds.

If you may't create suction with the plunger after forcing sufficient water out of the bowl, fill the bowl with a regulated quantity of water thru elevating up the flush valve in the reservoir. Plunge yet again.

When you're positive the drain is obvious, attempt a moderate flush, retaining your hand prepared to block the flush valve if the water threatens to overflow. After the obstruction has cleared, flush the relaxation room with a 5-gallon bucket of water to dislodge any very last materials.

Now which you realise a manner to repair the maximum common toilet issues, allow's attention on solving some of the most common bathtub and bathe problems.

Fixing Bathroom Shower and Tub Problems

Nothing feels better than coming home to a warm bathtub after a long workday. However, you can not experience this sort of clean satisfaction if you have a malfunctioning shower or bathtub.

Let's take a look at a number of the maximum common tub and shower troubles and the way you could repair them with out calling a plumber.

Repairing a Leaky 3 Handle Tub or Shower Faucet

If your 3-cope with tub and shower taps are leaking, don't trouble calling a plumber because that is a interest that almost all people can do. Like each different tap, a 3-manage tap can also leak because of normal placed on and tear or virtually growing vintage.

This type of faucet uses separate handles for bloodless and warm water, with a manage at the center diverting water glide from the tub faucet to the shower. You have as a manner to repair a leaking 3 control bathroom tap inside the following few steps, notwithstanding its complex look:

First, turn off the water supply to the faucet: Before starting your repair, make sure you have your water deliver became off. If your tap does now not have any close to-offs for the relaxation room, turn off the building's number one supply water valves. Drain the taps through turning the bloodless and hot tap knobs to the open function after turning off the water deliver. You can drain any ultimate water by using organising any taps located in decrease positions. This will ensure you don't have even a unmarried drop of water flowing for your defective tap.

Uninstall your faucet to appearance the trouble: Unplug the screws from the handles that manage bloodless and warm water. Pull the handles far from the tap stems with care. If the handles appearance rusty and are difficult to transport, lubricate them the use of oil. If the handles live stuck to the stem, detach them with any manage puller to keep away from destroying them. Attach the pulling tool to the faucet manage and twist it until the stem releases the deal with, turning anticlockwise to do away with the escutcheon panel in the back of the cope with—if it has one. If the escutcheon attaches to the plate via the usage of a screw, detach the screw and extract the escutcheon. With a flathead screwdriver, gently eliminate the escutcheon from the wall, exercising care no longer to deform it if it's far stuck.

Replace any worn-out elements: Turn the faucet address anticlockwise with a bathtub valve socket wrench. Remove every valve stem's rubber and screw washing system and tighten the screw after adding the contemporary washing machine—use a flashlight to observe the valve. If the tap seats are correct as new, there can be no want to repair them. However, you'll want to update the seats if they may be wiped out or scratched. Insert a seat wrench through every aspect of the valve and crank anticlockwise to put off the seats. Replace them with new ones and twist them clockwise to tighten them.

Return your faucets and check for leaks: To reinstall your taps, start through turning the stem clockwise into the valves, then fasten them with a bathtub socket wrench. After that, reinstall the faucet handles, which should be agency but not too tight. Reinstall the escutcheon plates and tighten the three tap handles. Turn in your water supply to look if there are any leaks. If you don't see any leaks, move in advance and take a tub or a tub and reward your self for a challenge properly executed.

Repairing a Faulty Shower Diverter Valve

The paintings of a bath diverter valve is to redirect water glide from the tub to the showerhead and vice versa. After use for a long term, this valve is at risk of located on and tear, inflicting it to malfunction or begin leaking. To repair this problem, follow these easy steps:

Shut off the water supply and uninstall the shower diverter valve

Before detaching the handles, ensure the water deliver to the tap is off. Using a screwdriver, pry the ornamental pinnacle from the valve manage. If the escutcheon receives within the way of using your deep-socket wrench, it may be vital to get rid of it. The Shower Valve Socket ought to be threaded over the diverting valve stem and tightened over the bonnet screw. Go earlier and uninstall it.

Inspect the seat washing device and screw

To monitor the screw and seat washing machine, pull the stem faraway from the relaxation room wall. Examine the inner component of the tube to ensure that no O-earrings or washers from the vintage valve are getting caught there. If you discover any, pull them out and throw them away. They might be the supply of all your diverter valve issues.

Replace the seat washing machine if want be: If you have were given a worn-out seat washing gadget, update it with a latest one blanketed with warmth-resistant faucet grease. Use the perfect seat washing machine duration and form, then press it tightly into the stem's retainer.

Reassemble the valve and test: You can reassemble the shower knobs and take a look at if the diverter valve is jogging nicely.

Chapter 5: Fixing A Clogged Shower Head

It may be very irritating while you are sincerely in search of to revel in a chilly shower, but the water is slightly popping out of your showerhead. Don't worry. This problem is easy to repair in just a few steps listed below:

Clean the showerhead

Sometimes, your showerhead has low water float because of clogging. That method you want to easy it and ensure all its nozzles are open. You can try this through rubbing the nozzles with a toothbrush.

Alternatively, wrap a plastic bag full of water mixed with vinegar on the ratio of one:1 for the duration of the showerhead, as shown within the photo above, to easy it.

If that doesn't paintings, unplug the showerhead the use of your hands or an adjustable wrench and extract the clean out show.

Clean the filter out display: Normally, the clear out display is a steel mesh disc protected with a rubber gasket and is placed at the aspect in which the showerhead connects with the water supply pipe. Its reason is to easy out any dust in advance than the water receives to the showerhead. Remove the display display screen with tweezers and smooth it very well.

Reinstall the showerhead: After cleaning everything thoroughly, positioned once more the showerhead with the beneficial resource of hand. You can use an adjustable wrench to tighten it on the joint. Turn on the water and check to appearance if the water drift is right.

Repairing a Shower that Won't Turn Off Completely

After using your shower usually, you could reach a thing in which water is left leaking from the bathe even after you switch it off. This genuinely technique that your shower valve washing device seat has malfunctioned. To restore it, comply with those steps:

Turn off the water supply: Start with the aid of way of turning any water supply to the bathe. Depending on your home's plumbing machine design, you may use the near-off valve inside the relaxation room or the primary house near-off.

Remove the faucet deal with: In the slender spot along the rim of the tap deal with pushbutton, insert the blade of a flat-headed screwdriver. Remove the button from the deal with with the screwdriver to show the screw. Pull the relaxation room tap address a ways from the shower stem thru doing away with the exposed screw. Pull the stem from the wall through turning it anticlockwise with a bathtub stem socket.

Extract the worn-out valve washing device seat: Simply input your seat wrench into the stem aperture. Remove the damaged valve washer seat through turning the wrench anticlockwise.

Install a brand new valve washing machine seat: First, exercise a lubricant to the current-day seat's threads in advance than placing them, then insert the seat of the wrench into the hole inside the stem. Tighten the replacement valve washing machine seat with the resource of way of turning the wrench clockwise.

Put the whole thing again and check: Place the stem back within the bathe wall's stem beginning. Tighten the bathe stem with the resource of turning it clockwise the usage of the bathe stem socket. Use the address to cowl the stem and join it with the nuts you took out in advance than. Reattach the button to the deal with and turn on the water supply to your bathe. Your bathe must now near well.

Fixing a Hand Shower That Leaks After Turning On Water

A leaking bathe head is generally sincere to repair. However, in advance than you start, you need to first decide the inspiration of the trouble, which is simple sufficient.

If your showerhead leaks after turning at the water, the rubber washing machine or the O-ring in the showerhead wherein it attaches to the hose is maximum likely responsible. If it leaks even if the water is have become off, the trouble is maximum in all likelihood with the cartridge hidden within the again of the cope with at the wall.

First, permit's check a manner to repair a hand shower that simplest leaks at the identical time because the water is strolling. Simply follow those clean steps:

Detach the hand shower: You most likely have a malfunctioning washing machine if your bathe head leaks from the intersection in which the supply line feeds into the hand shower head. Pull the showerhead off with the aid of manner of manner of pressing the discharge button. However, in case your showerhead does now not have a release button, actually rotate the nut protective the showerhead the usage of a bit of material for a first rate grip.

Remove the antique washing device and update it with a new one: You will see a rubber washing gadget in the the front of the showerhead. Remove it and update it with a state-of-the-art one.

Reassemble the showerhead and location it lower decrease again to test.

If the leak come to be because of a worn-out washing device, your shower need to now be as true as new.

Fixing a Hand Shower that Leaks With The Water Turned Off

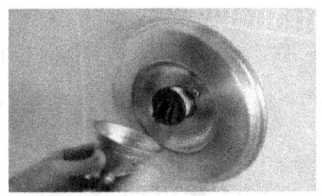

If your hand shower leaks despite the fact that the primary shower is grew to emerge as off, the cartridge is defective. The cartridge is a regulating device in the back of the cope with liable for setting up and last off the water deliver to the hand bathe head. To repair the problem, update this cartridge with a modern-day-day one thru the use of following the ones few steps:

Shut off the water deliver: You can do this from the bathroom or the house's fundamental supply.

Remove the take care of: Use a screwdriver to loosen the cope with screws and cast off the whole manage. Remember to place the drain stopper without delay to make sure that if any screws fall at the floor, they'll now not roll down the drain.

Extract the Cartridge: Now that the address is off, the cartridge ought to be visible. It is usually a pin coming out of a spherical assembly. Remove it lightly, but if it does not come out with out troubles, use channel locks or pliers to curve it out. Remove any dust or debris in that area.

Insert the state-of-the-art cartridge and check: Take your new cartridge and area it with the pin looking again similar to the vintage cartridge were positioned. Reassemble the manipulate and turn on the water supply for finding out. You ought to now have your hand bathe well repaired with zero leaks.

Now that we have finished the most not unusual bathtub and bathe protection, permit's go with the flow on to the subsequent economic catastrophe and look at the manner you want to restore your sprayer plumbing problems.

Repairing Faulty Kitchen Taps, Faucets, and Sprayers

Everyone likes to have a completely sensible kitchen—that's why we put money into sprayers to make cleaning and specific kitchen chores simpler.

Unfortunately, those sprayers are very prone to put on and tear because of the steady movement wanted even as the use of them. For that purpose, they have a tendency to preserve malfunctioning.

Let's have a take a look at how you could restore some of your sprayer troubles with out paying a plumber.

Repairing a Pullout Faucet That Won't Retract

When your pullout sprayer refuses to retract, you'll want to push it lower again physical, a tedious and unsightly undertaking. Pullout hoses have a further weight related to them, inflicting them to retract due to gravity. Therefore, after they fail to retract, it can be that this weight is off or that the spraying hose is vintage. To restoration the trouble, comply with the stairs under:

Check if the hose remains linked to the tap: Look below the sink to peer if the sprayer hose notwithstanding the fact that connects to the faucet. If the hose is separated, you may consequences reposition it by transferring the burden up or down till you have were given the hose correctly reinstalled.

Untangle the hose: If you discover your tap's spraying hose or tubing has emerge as tangled with the water tube valve, you can untangle it by way of manner of the usage of rotating the valve down and up. This may also moreover help the tap in beginning to retract.

Fix the Weight

If the problem is your tap's weight, virtually purchase a few faucet weight and update it using a screwdriver.

Repairing a Clogged Diverter Valve

A diverter valve allows water float in your sprayer after turning to your faucet. Unfortunately, after using the sprayer for lengthy, the diverter valve gets clogged with water minerals and one-of-a-kind debris blockading water from getting to the sprayer. To restore this hassle, study those clean steps:

Turn the water off and unscrew the deal with: Turn off the water valves earlier than the use of a flathead screwdriver to unscrew the number one deal with screw. After that, flip the cap anticlockwise to show the faucet's cap and detach the address.

Detach the spout: Remove the uncovered cam to benefit get proper of access to to the tap ball meeting, that you need to moreover take away. Then detach the spout by means of way of rotating it severally until it becomes free and is derived off.

Remove the diverter, clean it, and positioned everything once more: Locate the diverter within the the the front give up of the faucet's control and unscrew it so that you can also clear away the blocked dirt debris with vinegar. After washing and drying the diverter, pass back it to its specific region within the kitchen tap. Put everything decrease lower back because it changed into and enjoy using an unclogged sprayer.

Fixing Low Water Pressure in the kitchen

While low water strain has many reasons, a clogged aerator is the most common purpose, which regularly happens while you operate hard water inside the kitchen—it leaves at the back of mineral deposits that clog the aerator. Fortunately, this trouble is simple to repair if you test the steps under:

Open the spout to access the aerator: You'll want to untwist the aerator from the facet of the precept vent using your palms to transport it in a clockwise course. If the aerator will become caught and is tough to reveal, use a fixed of pliers to pry it open. However, take care no longer to scratch the floor with a employer maintain on the pliers by way of manner of wrapping protecting tape throughout the pliers' jaws. Having opened the spout, you could use your finger or a screwdriver to test if any aerator additives are stuck indoors it.

Remove any caught debris and minerals: Using your screwdriver, pry out any materials trapped in the spout, then easy away any closing debris. Using any tiny pointy device, disassemble every phase of your aerator. Before disassembling any element, make a highbrow be privy to techniques where everything is going. You can take images to apply as a reference later.

Clean each a part of the aerator in vinegar in advance than placing the whole thing back: Clean all the additives with the beneficial aid of dipping them in vinegar or wiping them down with a wet dipped-in-vinegar material. This aids in the dissolution of all mineral formations. Finally, rotate the aerator anticlockwise to rinse, reassemble, and reattach it to the spout. Make sure it's as tight as you could get it in conjunction with your palms.

Chapter 6: Fixing A Leaky Kitchen Tap Or Sprayer

If the pinnacle of your kitchen tap or sprayer is leaky, you haven't any specific choice but to update it. Luckily, it is an clean project that you can entire via following the smooth steps beneath:

Start through reducing off the water supply for your kitchen sink first: You can do this via turning off the water valves decided below the sink.

Remove the faucet head collectively with the C-clip: After lowering off the water supply, use your pocket screwdriver to do away with the vintage tap head, exposing the C-clip and casting off it.

Install a brand new faucet head: You'll want to put in the new tap head after eliminating the vintage C-clip. First, use a showering machine to regular the contemporary C-clip. After that, you may use your flat head screwdriver to place within the contemporary head thru setting it like your antique tap head.

Always hold your faucet head clean: After becoming a ultra-modern-day faucet head, unscrew it every so often to smooth out the dirt accumulated on it. That way, you'll decorate its common normal performance and decrease the chances of unexpected leaks.

Let's now have a observe repairing not unusual sink and faucet issues:

Sinks and Outdoor Faucet Repairs

Another set of not unusual plumbing troubles in a domestic revolves round sinks, faucets, and everyday faucets. You can also moreover want to have sinks and faucets in particular elements of your house, which consist of the kitchen, toilet, eating place, outside regions, and so forth. The extra sinks and faucets you have got were given, the greater plumbing problems you'll face.

Let's check how you can do some of those renovation.

How to Fix a Loose Sink

After the use of your sinks for a long time, you may word some of them feeling loose. They begin to shake and wobble on every occasion you use them. To make your sink business agency yet again, have a look at the steps below:

Inspect to appearance if the sink is in suitable circumstance: Sometimes, a sink starts offevolved to sense loose after breaking. In this form of case, you'll haven't any preference however to replace it with a modern-day one. However, in case you study the sink and discover that it's miles in pinnacle condition, you could only need to do some preservation to repair your hassle.

Look without spending a dime screws and attachment hardware: Sinks commonly connect to the wall the usage of screws and one in every of a kind attachment hardware.

If you find out that the screws are free, use a screwdriver to tighten them. Remember, all of it relies upon at the sort of sink you've got and the manner it changed into installation. If steel rods manual your sink, test to look if the steel rods are free or rusted, and update them to make your sink organisation another time.

How to Fix a Leaky Sink

First, have a examine the photograph beneath and be aware the exquisite factors of a sink to understand the repair technique higher.

Dealing with a leaky sink may be disturbing because of the reality your flooring will continuously be wet from the leaks. Luckily that is a problem you may recuperation thru following some steps indexed under:

Identify the leak deliver: Below each sink are numerous pipe combinations that drain water to your vital drainage line. Most sinks leak on the nut strainer, gaskets, or supply line pipes.

Replace any worn-out additives: If your sink is leaking from the location across the friction gasket, you have got a worn-out element. Use a spanner to disassemble the factors and update the tired elements with new ones.

Seal or replace any damaged pipes: As you could see from the photograph above, many pipes are underneath the sink. If your leak is at any of the pipes, the pipe desires sealing or alternative. Use a plumber's tape to seal any holes in the pipe. However, if that doesn't paintings, replace the broken pipe.

Reassemble the whole lot lower back collectively and test: After repairing, reassemble all of the additives and make sure each part is tight. Run water down the sink; you need to see no leaks.

Fixing a Clogged Kitchen Sink

While any sink is liable to clogging, the kitchen sink is the largest wrongdoer, in all likelihood due to the truth it's miles the most used sink in a domestic. When you phrase that water keeps flooding your sink, it's far a smooth signal that your sink is clogged.

Here are some clean techniques to recuperation the problem:

Pour warmness soapy water into the sink: Sometimes, the clog for your kitchen sink may be due to the fact you poured oily liquids down its drain. For instance, when you have been cleansing oily utensils on the sink, it can be that some of the oils iced up and clogged your sink. Try melting away the fat by means of the use of pouring warmness soapy water, and watch your sink unclog in seconds.

Clean the use of baking soda and vinegar: If heat soapy water does now not do the trick, attempt the use of a aggregate of baking soda and vinegar. This combination is same to the chemical materials plumbers use to unclog sinks and pipes. Use a cup or any place to cast off all the stagnant water from the sink. Pour about a cup of baking soda down your drain, ensuring you push the powder down the usage of a spoon if essential. After that, pour one cup of white vinegar down the drain. Close the drain via placing a stopper to cowl it. Give the mixture 15 minutes to do the magic. Remove the cap and flush the drain with warm water. This need to unclog your sink.

Use a plunger to unclog the sink: Another smooth manner to unclog any sink is to use a plunger. Simply vicinity it on the draining hollow even as the sink has a few water and plunge until it turns into unclogged.

Clean the P-trap: Do you take a look at the elbow-normal pipe below your kitchen sink? That is the p-entice. If the strategies above do no longer artwork and your sink remains clogged, your p-lure is probably the problem. You have to smooth it and do away with any debris that would be inflicting the blockage. Simply twist the round component to open up the p-trap. Remember, it is sure to get messy; be armed with some towels, gloves, and goggles. In addition, vicinity a bucket below the p-entice to make certain any water splashes into the bucket in desire to the floor. After casting off the p-trap, clean out any caught particles and rinse it with smooth water. Assemble everything decrease decrease lower back and enjoy your unclogged sink.

Repairing a Leaky Outdoor Faucet

Outdoor water taps that leak are a nuisance, especially at the same time as it's hot outside and also you need to pinnacle off your swimming pool or run your handheld sprinkler. In addition, water leaks are pricey ultimately, so it's crucial to restore any troubles along side your outdoor water faucets as fast as feasible. Fortunately, maximum tiny leaks are easy to restore, even for beginners. Simply have a look at the stairs beneath:

Start thru slicing off the water supply: Before repairing the outside faucet, endure in mind to show off the water supply truely. Otherwise, be prepared to get soaked. You will usually find the faucet shutdown valve inside the utility room, move slowly area, or basement.

See if the packing nut is loose and tighten it: After turning off the water, cross out of doors and unhook the hose out of your outdoor water tap. You have to find out the packing nut beneath the faucet's manage. This nut is typically accountable for offering a watertight seal across the valve stem. A leaking water tap can now and again be due to a loose packing nut. If that's the case, tighten it properly using a wrench.

Remove the valve and replace the bathing gadget if the leak is continual: If the leak is continual, get rid of the packing nut to get proper of get entry to to the valve and cast off it. Often, all you need is a stable draw close and a piece strive to drag it out. You can also take the faucet and the supplying pipe to your palms and flip them clockwise at the same time as dragging them upward. With a flat screwdriver, pry away the metal washing device at the base of the extended valve stem. Then, replace that washing machine with a cutting-edge one of the same period and reinstall the valve cease. Put the whole lot lower returned collectively and transfer the water all over again directly to affirm in case your faucet is constant.

Fixing a Frozen Outdoor Faucet

A frozen outside water faucet can motive catastrophic plumbing issues. When a faucet freezes, it exerts a full-size amount of strain, which could cause spigot additives to break and pipes to burst.

Luckily, with these few steps, you may preserve an outside tap from a spring freeze:

Open the tap and wrap the deal with with rags: After beginning the tap, wrap the address, spindles, and supply pipe in vintage rags or napkins. Make the wrap as tight as viable, but allow sufficient area round the faucet starting off for thawing water to go with the flow freely.

Pour warmth water on the wrapped faucet to unfreeze it: Slowly pour water over the blanketed faucet. Soak the towels in hot water slowly, then prevent and look for trickles coming from the tap. It can take some tries in advance than the spigot begins offevolved offevolved to unfreeze. Allow the water to glide for several mins after you have got a ordinary glide. Turn the tap off for a few seconds earlier than turning it once more on. Any frozen pipe additives in the lower back of the spigot want to melt as water flows through the spigot. Repeat the ones steps slowly to unfreeze your faucet.

Now which you apprehend a way to restore indoor and out of doors faucet problems, permit's glide without delay to drainage plumbing repairs.

Drainage Repairs for Tubs, Showers, Floor Drains, and More

As a home owner, the final component you have to underestimate is the significance of drainage troubles. While a clogged or slow drain can also moreover appear to be a minor hassle, it is able to rapid grow to be a wonderful problem.

Slow drains can sign numerous troubles, together with the early degrees of a blockage, narrower pipes due to mineral or fats accumulation, or maybe lousy drainage gadget grading. Whatever the motive of the trouble is, you want to repair it as speedy as viable.

Let's test how you can restore some of the maximum not unusual drainage troubles in a domestic.

How to Unblock a Clogged Shower/Tub Drain

If you word that water isn't draining out of your rest room floor or the tub after a bathtub, your bathe/bathtub drain is clogged. Follow the steps under to repair the problem:

Pour unclogging chemical compounds down the drain: Sometimes, the clog is due to soapy debris and hairs that you could easy without setting up up the drain. Buy an unclogging chemical from the hardware and pour it down your drain. Let it rest for 20 minutes, and then pour water. If the clog became not some thing severe, your bath and rest room floor must now drain with out issue.

Remove seen clogs via manner of hand: A blockage can sometimes be visible on the pinnacle and is straightforward to attain. Put on a couple of latex gloves and try and cast off the blockage as thoroughly as feasible. Alternatively, you could use small plastic hooks to reach the clog. Although it may be charming to use a coat hanger in this situation, be aware that doing so also can motive harm on your drain.

Use a drain snake for clogs too an extended manner down the drain: A drain snake or a plumber's snake is stretchy and extendable drilling device to be had at hardware shops. You'll shoot a steel twine into your drain the usage of the snake's hand crank to split up or pull out the entirety blockading it. There are also disposable plastic snakes available that don't require any cranking and may even be used to unclog a toilet. Remember to smooth your drain snake whenever you use it.

Pour heat water down the drain: If there aren't any sizable impediments obstructing the drain and the use of a plumber's snake hasn't labored, cleansing the drain with warm water can help disintegrate softer buildup from spherical the edges—like cleaning cleaning soap scum. Repeat the machine or 3 times extra if important.

Chapter 7: Repairing A Broken Underground Drainage Pipe

A broken drainage pipe for your garden won't appear to be a severe hassle, but it is. It can purpose flooding, contaminate the soil and water, and reason many health troubles. Fortunately, solving the trouble is easy due to the truth all you need is to comply with the stairs below:

Locate the pipe breakage and dig spherical that area: To begin, dig a hollow in the earth to find the broken pipe, so that you can maximum likely be near any visible damage signs and symptoms, at the side of moist spots. Dig an extra six inches below the broken pipe vicinity after finding the broken pipe area to create a strolling location for yourself.

Cut out the damaged element: You'll want to reduce out the broken area of the pipe to restore it. Measure inches on each aspect of the damaged spot and label the pipe earlier than decreasing. Labeling is vital for accuracy features.

Fix a brand new piece of pipe: Take measurements of the broken pipe you eliminated in advance than including a today's PVC pipe segment to make sure the pipes combination in. The new pipe have to have the identical thickness due to the reality the old pipe to save you leaking. After that, attach the modern day PVC pipe to the modern-day pipe ends with bendy rubber connectors. Then slide the elastic connectors to cowl every the vintage and new pipe sections.

Confirm that there may be no leakage and cowl the floor: After converting the damaged element, allow a few water run via the drain to affirm no leaking. After that, shovel lower lower back the soil you had dug out to cowl the hollow.

How to Repair a Broken Shower Drain in a Concrete Slab

If you have got an upstairs rest room, the bathe drain probably passes down behind your walls. If you observe wet patches on that part of the concrete wall, it could endorse that the drain is broken.

Here is the way to restore that trouble:

Remove the concrete throughout the drain: To dig out the concrete throughout the drain, you could need a hammer and a chisel. Dig out the concrete till you note the bolts and flanges maintaining the drain line towards the wall.

Unscrew the drain line from the wall and dispose of it: Using a wrench, take away the screws keeping the drain line in place and dispose of the vintage damaged drain

Install a modern-day drain: Replace the antique drain with a contemporary one. However, make certain that it's far the equal period to ensure it connects effortlessly with the alternative drain components. After that, check for leaks by using manner of using pouring water into the relaxation room drain.

Put lower back new concrete

After solving the modern day drain line and screwing it once more to the wall, prepare new concrete and plaster it at the wall to cowl the drain. Paint the brand new concrete to convey your wall decrease returned to its unique look.

How to Fix a Smelly Floor Drain

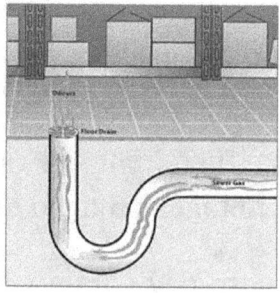

Most houses have floor drains inside the bathroom, balcony, and one in every of a type additives that need to empty water. Sometimes, you can word a horrible scent coming from the drain hollow. To repair this trouble, observe the stairs underneath:

Flush warmness water down the drain: Sometimes, the stench out of your floor drain results from oily debris clogged within the pipes. To wash this particles away, pour warm water into the drain.

Pour a mixture of vinegar and baking soda down the drain: This trick works like unclogging kitchen pipes. The vinegar and baking soda dismantle all of the clogs and dispose of the horrific scent.

How to Repair a Leaky Plunger-kind Drain on Your Tub

If your tub comes with a plunger, it most likely does no longer have a seen stopper. Instead, the plunger's lever pulls the hole metal plunger down and as an entire lot as seal the opening. Therefore, if water seeps down the drain, the plunger is broken or clogged. If cleaning doesn't recuperation the trouble, test the steps below to replace the plunger and join the trouble:

Extract the antique plunger: Use a screwdriver to unplug the screw that secures the experience lever plate. Lift the linkages and metallic plunger out of the overflowing drain thru raising the ride lever plate. If the plunger does now not connect to the linkage, insert a bendable grabber into the hollow to extract it.

Install a brand new plunger: Assemble the modern plunger and installation it inside the same role due to the fact the antique plunger. Screw it in area and check to appearance if water remains seeping thru. If water seeps via, modify the length of the linkage till it's far tight sufficient to save you water seepage.

I want that with the resource of now, you may repair severa drainage troubles round your private home. Let's circulate on and be aware how you can manipulate commonplace piping issues.

Handling Pipe Repairs

You already realize that pipes are a primary part of your property's plumbing device — through pipes, clean water receives into your house, and the grimy water gets out. Given this, expertise the way to restore precise forms of pipes round your property is a skills each property proprietor need to have.

Let's dive proper in.

How to Repair a Leaky PVC Pipe Using Fiberglass Resin Cloth or Tape

If you phrase a PVC pipe spherical your home is leaking, and you want a short answer, fiberglass resin tape is your excellent guess. Simply observe the steps under to seal the leak:

Identify the leaking factor pipe: Even despite the fact that you could locate water throughout the pipe, the possibilities are that the leak is handiest on a small a part of the pipe. Dry the pipe and perceive the ideal vicinity of the harm.

Place the tape at the broken detail: Use a moist rag to clean the damaged segment of the PVC pipe earlier than utilizing the tape. Next, wrap the tape band throughout the broken region earlier than the resin dries/treatment plans then allow about 15 mins for it to treatment.

Test to make certain there can be no more leakage: After searching out 15 minutes, the leakage need to end— the fiberglass tape inhibits the leakage through the usage of a water-activated glue, which commonly hardens across the pipe.

Alternatively, use a fiberglass resin fabric as opposed to tape: Replace the tape with a fiberglass resin material if you want a extra everlasting answer. First, clean the damaged thing and use sandpaper to rub the place to make it greater adhesive. After that, cowl the damaged detail the usage of a resin material and positioned a UV lamp straight away at the pipe or region the pipe piece below direct sunshine. This will treatment the resin fabric and make it stay with the pipe eliminating the leak.

Repairing PVC Pipes Using Epoxy

Another manner to restoration a leak on any PVC pipe with out replacing it's far to use epoxy using the steps under:

Identify the leaking detail: Epoxy is a repair fluid that hardens in mins. To use, you ought to first identify all of the leaky factors of a pipe earlier than beginning to use the liquid. Wrap a dry towel throughout the regions you suspect to be broken and if the towel turns into wet, mark the ones areas with an outstanding pen.

Turn off the water supply: After identifying all of the damaged additives, turn off the water deliver. That way, you could have a dry pipe with which to art work.

Make the pipe ground hard: Because the smooth floor of PVC pipes makes adhesion complicated, you'll want to hard up the pipe with popular-sized sandpaper to make certain that the epoxy adheres properly. After sanding, use a smooth, dry cloth to wipe away any dust.

Mix the epoxy: Mix the epoxy components in line with the ratio indicated via the producer. These ratios range from brand to logo; observe the producer's manual. Remember to use gloves on the same time as mixing the epoxy additives. More importantly, most effective combination epoxy that you can use in three minutes because it dries up very speedy. Therefore, if you have several or many leaks to seal, you will likely need to replicate this mixing step severally.

Apply the blended epoxy: Wrap the very last epoxy round your pipe in the marked spots. Make certain you press it on and amplify it through at least an inch on all of the broken

components.

Let the epoxy treatment and test for leaks: Although the epoxy plaster remedies in about ten minutes, it takes about an hour to set completely. As a stop result, you have to provide it at least this extended without turning on the water supply. Apply the hand towels take a look at to the repaired components to ensure that the restore has addressed any leaks. Although epoxy putty will offer a long-lasting restore, it's still useful to test on it periodically in a while to ensure your restore labored.

How to Replace a Part of a PVC Pipe

Sometimes, a part of your pipe may be too damaged to restore the use of epoxy or distinct materials. This leaves you and not using a choice however to replace the part of the critically broken pipe. Follow the steps under to carry out this assignment:

Cut off the water supply first: Shut off the water deliver first and permit all the water drain away.

Cut out the damaged a part of the PVC pipe: Use a hacksaw and decrease out the broken part of the pipe. Measure at the least 2 inches on both elements after the damaged element and make your cut there.

Cut a piece of the modern pipe and make it equal to the antique one.: Cut a bit of PVC pipe the identical diameter because the old pipe to the volume of the damaged detail. Use sandpaper to smoothen the ends of the contemporary pipe and dispose of any burrs.

Combine the vintage and new pipe: Spray PVC primer and cement on both the current and new pipe ends, after which be a part of the 2 with a coupler. Slide the rims of the modern-day pipe into the winning pipe, then rotate the joints a hint to disperse the mortar in among PVC joints. Give the repaired pipe 15-1/2-hour to dry in advance than the usage of it.

Turn on the water deliver and check for leaks: After the primer and cement are dry, permit water go with the flow thru the pipe and check for any leaks. If you measured the entirety flawlessly, there ought to be 0 leaks.

Repairing a Burst Metal Pipe Using a Pipe Patch Kit

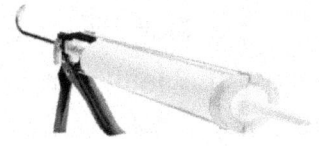

Having a burst metal pipe spherical your house can be annoying because it regularly approach you need to update it, which may be pricey—no matter the whole lot, you will want to buy a contemporary pipe. However, as you prepare to replace the pipe, you could repair it the usage of a pipe patch kit to save you in addition leaks. That way, your private home will no longer be bothered with the aid of water harm. Simply examine the steps under:

Turn off the water deliver: You can do this at the precept water transfer. In addition, open all taps above the pipe to take away any water left in the pipe device.

Inspect and dry the leaking place: Take a smooth rag and wipe the burst region of the pipe. Scrub the outdoor of the pipe with steel wool if there can be any lime or rust scale, then have a look at the supply of the leak. If a pipe junction is leaking, use an adjustable wrench to adjust the joint turning into until it's miles tight enough. After that, mark all the broken factors of the pipe.

Chapter 8: Why Plumbing Maintenance Is Important

Maintenance of any kind is crucial, which includes upkeep of your property's plumbing device. This technique making sure the lavatories, sinks, bathe, tub, or distinct gadgets associated with plumbing artwork efficaciously.

Repairs need to be completed to your plumbing systems earlier than the trouble escalates. Ensuring everything is because it should be strolling has large benefits which incorporates:

1. Reduced Water Bills

Did a drop of water leaking each second can translate to 3000 gallons of water each 365 days? Shocking, is not it? Well, that waste will increase your water bill. That is why you should now not neglect about approximately any leakage, no matter how small.

2. Improved Water Quality

As soon as you phrase rust for your water pipes or the water, restore it right now. Rust is a water contaminant and isn't appropriate to your fitness. Besides, a leaking pipe without trouble enables micro organism increase, which includes giardia, legionella, or cryptosporidium. A leak diagnosed early will save you micro organism from growing, making sure water is simple and wholesome for use.

3. Better Quality Of Air

Faulty pipes and cracked pipes that lead to water leaks can affect the best of the air you breathe in your home. Mold is probably to increase, growing stale, musty air that could create undue tension on the heating, air waft, and air con tool. It will astonish you at how brief maintenance to leaking pipes can actually get rid of the mildew.

four. Longer Life For Your Plumbing System

The functionality of your plumbing device is maintained at the same time as problems are regular proper away. For instance, high water stress want to be decreased with on the spot effect to keep away from placing a stress at the pipe joints. If left unattended, it may reason the pipes bursting.

To boom longevity on your water home gadget, attend to water leaks proper away to prevent water from collecting on the system's base, which motives incredible harm.

five. Better Health For Your Family

Water leaks purpose mould boom. Out of the 21.Eight million cases of bronchial hypersensitive reactions inside the US, a spectacular 4.6 million were associated with dampness that arises because of building up of mildew in a house. Proper upkeep can preserve bronchial asthma, allergic reactions, and awesome respiration illnesses at bay.

Water leaks cause wet construct-up inside your own home, which leads to mould increase. Of the 21.Eight million stated having allergic reactions inside the US, 4.6 million instances had been associated with dampness in addition to publicity to mold in houses. Plumber protection helps drastically preserve off such things like allergies, allergies and particular considered one of a type breathing situations.

6. Improved Home Value

Did you recognize that plumbing systems account for 10% of the house value? General protection will preserve the tool's function, thereby enhancing the price of your home.

Why You Need the Basic DIY Plumbing

You probably in no way notion approximately the network of pipes in your own home bringing in water and casting off waste till a hassle arose. So you contact a plumber to head returned and check what the trouble is and after that, pay a hefty quantity for a minor hassle that you could have with out difficulty steady your self.

This delivered approximately you to learn how to repair easy problems together with a rest room no longer flushing nicely, or a sink no longer draining water because of a blockage, or an disturbing water leak that you attempted to include through putting a bucket below to keep the water. So, why is DIY fundamental plumbing crucial?

1. Cost

Cost is one of the primary motives you'll determine to carry out easy plumbing safety. Plumbing problems will upward push up once in a while. Learning the way to repair minor troubles will save you cash in the long run.

2. Handling Emergency Repairs Effectively

Handling emergency safety is available in available as you may make short protection to healing a hassle as you plan to make the number one repair. For example, easy information at the side of understanding wherein the near-off valve for your water deliver is will considerably benefit in conditions wherein there can be a leak. This can prevent number one flooding in your own home.

Using the nice tool, you can remedy small plumbing troubles in your private by manner of learning some fundamental instructions.

What do You Need to Know Before Embarking On a DIY Plumbing Journey?

Just as with every studying software, you need to recognize and prepare yourself for what is earlier. So what's required of you? Are there sacrifices you could make want to?

It's critical to notice the following as you embark for your journey on DIY plumbing:

1. Cost of Learning

There are preliminary prices worried. This includes an eBook like this one with a manual on the manner to DIY plumb. You may even need to apprehend the primary-hand safety device you want earlier than embarking on a plumbing undertaking which I will cover in this e-book.

2. Your time

It requires a while to take a look at and exercise what you've got were given were given found out, which could purpose a hobby or an profits-producing aspect-hustle.

When Do You Call In A Professional?

As a good deal as DIY has critical blessings, there may be damaging outcomes as you try and restore them. A masses bigger problem can rise up, ensuing in better than everyday prices. For example, if the temperature and pressure have incorrect settings, it can harm your plumbing device. Using the wrong substances also can wreak havoc. More complex tasks ought to have bad factors outweighing the benefits. So which responsibilities are you capable of do thru your self, and which of them do you go away truly to a plumbing expert?

First, you need to apprehend the constructing codes for your location or area. Some codes require that a few plumbing jobs be tackled only with the aid of an authorized expert, even as others will can help you DIY.

Second, relying on the assignment, DIY plumbing may be charge-powerful while solving a minor trouble. You are taken into consideration a smart proprietor of a residence if you could make minor fixes or adjustments in your home. That method you may have to buy an ebook, check easy plumbing, and ultimately purchase additives and machine that will help you to your new hobby or alternate.

Third, tackling or managing DIY tasks outdoor your records or capability stage might also need to result in errors which can reason more problems in preference to fixing them. It might be notable to lease a plumber who will expertly resolve your plumbing troubles in such conditions.

Plumbing Safety Tools

Now that you have determined to analyze some simple plumbing abilties, the next issue you need to do is understand the plumbing gadget for you to preserve you stable as you determine for your initiatives.

Purchase all the crucial defensive equipment you could want, collectively with a warmth shield, gloves, boots, goggles, protecting helmet, and face defend. Sometimes, a plumber also can need respiratory gear to shield their breathing device.

1. Heat Shield Or Pad For Plumbing

Aad for plumbing will protect you towards burns and sparks while running on small welding, soldering, and brazing initiatives.

2. Goggles

Accidents can take vicinity while doing plumbing jobs, and prevention is commonly higher than remedy. Shield your eyes from flying wood chips, grease, splashes of chemical materials, grease, oil, fumes, and burns the usage of googles. It's smooth to forget and wipe a risky substance into your eye, causing grievous harm.

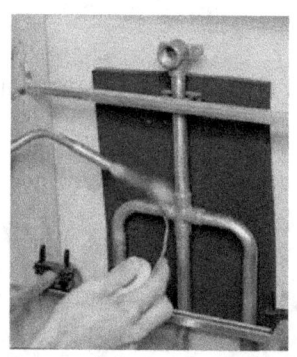

You will want multiple goggles that protect you from UV slight, commonly if you artwork outside for lengthy hours on a selected plumbing undertaking.

three. Gloves

You might imagine that there are not any risks on the same time as walking on a leaking tap or damaged flash, but having the proper defensive tools to your fingers, could make an massive distinction.

You will speedy discover that you'll be strolling in moist and slippery environments, and consequently gloves with an fantastic grip can save you any errors from occurring. Gloves can defend your pores and pores and skin from getting callus, which takes location due to abrasion while going for walks on metallic items.

four. Protective Helmet

This hat is crafted from hard fabric and is used in environments that might placed one liable to falling items. There are suspension bands inside the helmet, which spreads the weight and pressure of any impact struck on the pinnacle. These helmets may be used as protection from rain and electric powered powered powered surprise.

5. Boots

As you embark to your first plumbing job, you could discover that you will be walking in slippery and moist locations as you operate heavy tools. It's therefore critical which you get the proper boots on your type of interest. You can not select out simply any boot. You will want a pair this is durable, comfortable, and water-evidence.

6. Face Mask

You might also want a face mask, relying on what you're operating on. How well a tool is maintained can be a detail that determines if you'll put on a face mask or no longer. This is crucial to prevent publicity to dirt, and molds, which may be risky in your respiration.

Chapter 9: Understanding The Water System In Your Home

Now that you are ready to learn how to recuperation minor plumbing troubles in your private home, it is paramount that you apprehend how the water gadget in your house works. Being knowledgeable at the valves, pipes, and their components will offer you with a clean photo of the manner to solve issues and efficiently provide an reason for to a expert while extra top notch safety are required.

Despite their length or format, all homes have notable plumbing structures; one controls the input of water, and the possibility drains waste out of your house.

The Water Supply System

Water flows into our homes through being associated with the water deliver system. The following plumbing furnishings are the issue of touch from wherein we benefit the water to apply:

- Showers

- Bathtubs

- Toilets

- Kitchen sink

- Bathroom sink

- Faucets

- Any location that has strolling water

The gadget has numerous additives:

1. Pipes

Hidden below the flooring and at the back of the walls, water flows thru pipes that run within the course of your private home. Water stops moving whilst the taps are closed.

There is a remarkable pipe that connects to the valve thru a faucet, and you can near it off even as the want arises. However, it remains open at all times to allow water into your home. A plumber shuts it off whilst jogging on a repair, and you can need to try this too.

2. Water Heater

Your showers, bathtubs, or sinks are related to 2 pipes; one that brings in bloodless water, and the possibility brings in heat water. The cold water pipe is proper now related to the number one pipe, and the hot water pipe is set up to the precept pipe and continues to your heater, wherein the water is heated.

3. Shut-off Valves

These are valves that allow water into your property. When they're grew to grow to be off, water stops flowing altogether. That is specially accomplished when you want to paintings on a fixture, equipment, or pipe.

four. Faucets

The taps control the water that flows into your showers, sinks, or bathtubs. You manipulate the knobs whilst you need to apply water, as an example, out of your kitchen sink. All plumbing furniture have knobs, one for cold water and the other for hot water. When each are opened, warmness and bloodless water blend, growing a warm temperature.

If you need greater hot water, open the modern day water knob greater than the bloodless one, and vice versa.

five. Water Meter

The water you get hold of in your private home isn't loose. You pay for it. That is why before water is delivered to your property, it passes thru a water meter found internal your compound. It is installed to the primary supply pipe, and it indicates how a good buy water you have were given used, and then a water invoice is sent to you on the give up of each month

The Drainage System

Ever at a loss for words wherein the wastewater is going? It is worn-out away from your house through pipes linked to a treatment plant. When you flush your rest room, or take a bath, or drain your kitchen sink, all of the water flows outwards thru pipes in the drainage device. In the city regions, the grimy water drains away via sewer systems. In the rural areas, there are septic systems to eliminate the waste.

The additives in a drainage device include the subsequent:

1. Drain Pipes

Fixtures from your property have drainage pipes related to them that drain out wastewater. These pipes depend upon gravity as they're regular with a downward attitude to permit free flow of the waste and into an underground sewer pipe. The sewer line in the long run drains proper into a septic system built a long way faraway from your private home.

2. Drain Traps

Have you ever observed the U-normal pipe beneath your sink or toilet? Well, that is a drain entice. It is designed that manner to keep water constantly, to save you waste from flowing backward. Their shape allows water to constantly be held in the pipes to prevent gases in the sewer and ugly odors from moving into your home.

three. Drain Vent

Have you ever seen drain vents on the roofs of houses? These pipes allow air to flow into the drain pipes, putting in the right go together with the glide of wastewater within the drainage gadget.

Keeping the vent clean of any debris is critical as it prevents the backflow of waste.

RV Plumbing

The plumbing device to your RV lets in you to clean your dishes, shower, and carry out primary circle of relatives cleaning. Also, RVs are ready with toilets, until you are attempting to preserve up on place- you get to keep away from the usage of public toilets. The quality thing approximately your RV is that you could perform minor plumbing preservation honestly as you will in a residence.

How Plumbing In Your RV Works

The plumbing system in your private home and RV work the equal in the rest room, sink, and bathe. The critical difference is that your private home's plumbing tool is hooked up for your town or city's sewage system. Except for a leak in your own home, it in no way crosses your thoughts in which the grimy water goes, after flushing it, or as quick due to the fact it's miles going down the drain.

The RV plumbing gadget is self-contained. You need to empty it manually frequently.

Under the chassis of your RV, you will discover a tank that brings in freshwater. This is the water that flows in your faucets and showers while grew to become on. There's additionally a gray tank wherein waste from the sink and bathe drains into. The black tank holds water that contains waste from the rest room.

Gray water isn't very dirty and may be disposed of on the floor however need to be located via a gush of freshwater to save you tough scum from forming. On the opportunity hand, black water poses a fitness and environmental hazard and have to be associated with a sewer line while doing away with the waste.

Chapter 10: Diy Kitchen Repairs

Kitchen plumbing problems do now not just end up an inconvenience however can purpose good sized damage. While you'll require professional assist to restore a prime trouble like a broken pipe, a few are specifically simple to restore.

RV plumbing in your kitchen is just like that in our homes. The sinks are similar to what we use in, which incorporates the faucets.

The most not unusual problems within the kitchen plumbing that you could restore with the useful useful resource of your self embody the following.

- Clogged kitchen drains

- Not sufficient, or no heat water

- Dishwasher leaks

- Garbage disposal care

- Clogged or leaking kitchen sink

- Leaking kitchen tap

- Low water strain on the kitchen sink

Let's check them in detail:

How To Repair A Clogged Kitchen Drain

The kitchen drains clog due to grease and meals particles. Grease starts as a liquid, but as it flows down the drain, it hardens. You do not have to call a plumber while your kitchen drain clogs up. The following are 4 strategies to unclog your drain.

1. Boiling Water

Before the use of a plunger, one of the only techniques to unclog a drain is the usage of warm water.

Supplies favored

- Tea Kettle

- Water

The clog, specifically grease, can be dissolved through using the use of pouring boiling water down the drain. Hot water from the faucet will not do the pastime. It must be boiling water. A word of caution; do not pour boiling water proper right into a porcelain sink. Don't use it both in case your pipes are plastic because warm water can melt them. For the plastic pipes, use a plunger.

Method

1.Turn in your variety and boil a gallon of water in a teakettle.

2.Carry it over in your clogged sink.

three.Pour the new water down the drain slowly. If it does now not dissolve the clog, then you could choose out to apply a plunger.

2. Drain Plunger

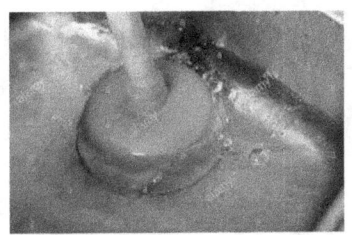

You will want the right plunger in your kitchen sink. A cup plunger is the most appropriate to apply.

Method

1.Fill your sink bowl with some inches of water. This manner, the cup of the plunger creates a respectable seal.

2.Place the plunger on the drain beginning. Make tremendous it seals properly.

3.By the use of quick plunges, pump the plunger up and down frequently. Do this typically, checking for the clog in order to be pumped out.

four.After non-prevent pumping and the clog has been eliminated, open the faucet and permit it run for one or mins to make sure the drain has cleared.

3. Sink Auger For Stubborn Stains

If the cup plunger does not unclog the drain, strive a sink auger.

Method

1.Remove the stopper

2.Extend the cable of the auger into the drain beginning. When it bumps right into a clog, pull out 12 inches of the cable and tighten the setscrew linked to the auger canisters.

3.Without using too much strain at the cable, turn the crank deal with clockwise. Turning the cable assists in breaking aside the clog or helping get past the bend. This prevents the cable from getting caught inside the bend or at the clog.

four. Continue extending the cable till you revel in you've got loosened the clog. This is specifically reachable if the clog is in the entice.

five.Pull out the cable from the pipe, pushing it lower once more into the canister. You also can see clogged depend stuck onto the stop because it comes out of the drain.

6. Repeat as frequently as desired until the clog clears up.

7. Immediately the drain starts offevolved to float, open the tap and allow the trendy water flush through the drain.

4. Remove The Drain Trap To Auger The Branch Drain

If you still can't gain the clog after extending the cable into the entice, it's miles probable within the branch drain or the vertical drain within the wall.

Supplies wished

- Bucket

- Sink Auger

- Wrench Pipe

Reaching clogs past the entice may want to require a bit extra paintings however nonetheless easy to do.

Method

1.Have a bucket underneath the lure to seize the water that flows out at the same time as the entice is removed.

2.Use tongue-and-groove pliers or a pipe wrench to unscrew the slip-nuts and the entice. You'll word a nylon slip washing machine firmly orientated into the pipe. Reassembling becomes easy.

three. Let the water pour from the trap and into the bucket. The disconnected ends of the pipe can even drain water into the bucket.

four.Push the auger cable horizontally in the drainpipe. Crank the cable as you tighten the setscrew, making use of stress to assist get past the bend.

five.As you parent the cable deep in the drain, you could art work via the clog.

6.Remove the cable after you feel the clog has cleared.

7.Put decrease back collectively with the drain trap. Open the tap and flush warm water down the drain to dispose of any debris final.

five. Should You Use Chemical Drain Cleaners?

Wouldn't or no longer it is plenty much less complicated to apply chemicals to unclog your kitchen drain? It may also look like the suitable possibility however do not chemically clean drains. They are not most effective an environmental risk but also are unstable in your kitchen pipes. Besides, they may be harmful for your health.

Also, chemical cleaners typically do no longer cast off the clogs within the drains. Instead, you'll be left with a sink complete of toxic chemical substances, and worse although, your sink can be entire of undrained water.

How To Fix an Overflowing Dishwasher

Your Dishwasher has continually faithfully carried out its manner, and also you paid little interest to it until a hassle arose whilst within the destiny you turn in your Dishwasher, and as it's far operating, you be aware an overflow of water.

Don't name in a plumber genuinely but. This is a hobby that a newbie can restoration.

A flow authentic like a saucer located within the basin rises because of the truth the water tiers in your dishwasher upward push. When the water reaches its most peak, the float signs your Dishwasher to reduce off the water supply. If the float becomes immobile, the water will hold to rise until it overflows. If you have got a dishwasher on your RV, you could recovery it the use of the guide.

Supplies desired

- Dishwasher drift transfer

- Flathead Screwdriver

Method

1. Unstick the Float Manually

Usually, solving a flow this is caught is easy. You only want to elevate and reduce it via hand some times to get it again to its feature. When the go along with the float is operating well, it actions freely to motive the transfer. If that does not take place or the waft is visibly broken, you need to update the entire go with the flow assembly.

2. Inspect the overfill go together with the go with the flow transfer

The float switch is smaller than a cigarette lighter, and it cuts off or allows a low-voltage electrical modern to waft. If you boom the go with the flow and pay attention a tough clicking metallic sound, then there can be an top notch chance it's far nevertheless in specific working circumstance. If you do not listen the sound, then the flow goals opportunity.

3. Replace the overfill flow switch

At this issue, you're very near fixing the dishwasher water overflow trouble. The next step is to get the guide for your Dishwasher and discover in which the go with the waft switch is. Different fashions have precise places; some are at the the front under the toe kick, on the identical time as others are all over again.

Before you replace the glide transfer, close down the electrical present day. Replacement is easy. Remove the damaged piece and fix the present day one. Once you're sure you have got got got steady it effectively, spark off the present day-day and test it all over again. If it's far working nicely, you may preserve to apply it like in advance than.

How to Clean a Garbage Disposal Sink

When your kitchen sink starts offevolved to provide off a awful scent, it's miles a first-rate signal that you have to smooth your drain and rubbish disposal.

The rubbish disposal typically cleans itself through way of way of breaking aside the meals particles and driving them down the drain, retaining the chamber specifically smooth. However, with time, pungent slime builds up inside the areas that do not get wiped clean through manner of manner of the scrubbing motion in the grinding chamber.

Cleaning it every week is a brilliant way to keep away the awful smells.

Supplies Needed

- Rubber gloves

- Sink stopper

- Dish cleansing cleaning cleaning soap

- A kitchen sponge with an abrasive issue

- ½ to one cup vinegar

- ½ cup baking soda

Other non-obligatory fresheners that you could use instead of vinegar consist of baking soda, rock salt, beach solution, or citrus peels.

Chapter 11: Diy Kitchen Installations

Just as you do not want to panic and get in touch with in a plumber on the equal time as plumbing upkeep are required, there are kitchen installations that you may do for your self, while a novice, in your house or RV. You can with out issues control putting in a dishwasher, changing a kitchen pipe, and putting in a kitchen tap.

Ensure to check if you require permits out of your nearby authority to carry out particular installations.

Installing A Dishwasher

Installing a dishwasher is simple, and also you do no longer need to put money into a plumber to do it for you.

Tools Needed

- Wrench

- Pliers

- Screwdriver

Materials wanted

- Dishwasher

- Wire nuts

- Screws

Method

1. Attach the drain line

Take your Dishwasher out of the sector, and lay it face all of the way all of the way all the way down to see its back. Double-test that every one the connections are intact.

Remove the cap from the drain line.

2. Thread The Drain Line

Crimp the clamp across the hose. You will want pliers to secure it properly. Thread the drain line into the hole below the sink.

3. Place the Dishwasher Under the Counter Top

Carefully location the new Dishwasher below your countertop.

Adjust the dishwasher legs' pinnacle via leveling the countertop the use of a wrench with the screws anchoring the Dishwasher to the countertop.

4. Reattach the Wires

Ensure the electricity is off on the breaker. Reattach the wires carefully, matching the colours, then screw at the wire nuts and sooner or later wrap with electric powered tape.

Put the copper ground twine below the inexperienced-spherical screw, and then tighten downwards.

5. Connect the Supply Line to the Dishwasher

Using a wrench, join the Dishwasher and the supply line and join the drain line to the plumbing system.

Turn at the water that is below the sink.

Connect the drain line to the plumbing device with the aid of the usage of tightening the clamp firmly across the hose. After making the connections correctly, flip the strength on on the breaker.

You can installation a dishwasher relying in your RV's length of. Nonetheless, it is ideal to comprehend that you may use lots of water, however you need to maintain as a whole lot water as possible. Because you want to frequently fill up your freshwater tank, you most possibly want it to very last for a long time.

Also, you'll use immoderate voltage power from your generator, which wishes to last for as long as you're on the road. If it's miles your own home, power conservation is a subject. Furthermore, you want to winterize it to save you freezing at some point of wintry weather.

How to Replace a Kitchen Pipe

Replacing a kitchen pipe may moreover look complicated, however it's miles one of the most effective duties to carry out, even for a amateur. You also can need to trade it because of the fact it's miles leaking or for a few distinct cause. The curved pipe is attached to the waste lines and the sink drain. As you prepare to repair new pipes, it's far proper to make certain that they're aligned. If the alignment is off even barely, it might not seal, and it will maintain leaking.

Replacing a kitchen pipe in your RV is the same as that during residential houses. The extraordinary difference is that the drain pipe connects to at least one sewer line in your home, at the identical time as within the RV, it connects to two separate tanks, the gray and the black tank.

1. Remove the Old Pipes

Removing the vintage pipes ought to not be difficult. However, you may want the crucial system to loosen the P-pipe that can't come off with out problem.

Supplies Needed

- Hacksaw

- Adjustable pliers

- Bucket

- Tape diploma

Method

1. To Begin...

Place a bucket below the curved part of the P-trap. Turn off the water valve to save you getting soaked if the faucet is open with out information on the same time as working at the pipes.

Inspect and take measurements the use of tape so you may moreover buy the right length. Using adjustable pliers, draw close the nuts located on the pipes extending down from the drain and the sink, and turn it clockwise. If the pipe is metallic, you could should spray lubricant to loosen the threads' nuts.

Disconnect the traps thru pulling them away and turn them the wrong way up to empty water into the bucket. If the traps have been linked to a commonplace tee, unscrew the relationship and dispose of the tee.

If crucial, get rid of the tailpieces. To remove them, unscrew them from the drain thru turning clockwise using adjustable pliers.

2. Install The New Sink Drain Pipe

Once you've got eliminated the vintage pipes, you are now geared up to put in the modern day pipes underneath the kitchen sink. To gain a really ideal connection, you need tight and particular fittings.

Screw-in the tailpieces; they are lengthy and may increase to the same diploma because the drain line's top. Extend the tailpieces to at least 2 toes above the cupboard's backside- if the drainpipe is at the floor. With a hacksaw, reduce the tailpieces to the favored duration.

Still the usage of a hacksaw, cut the ultra-modern pipes to healthy with the vintage ones. Ensure they may be the equal diameter. PVC pipes are favored in the kitchen because of the reality they do not rust easily.

Connect the pipes with compression fittings, after which use your hand to tighten the nuts up to the factor in which you are prepared to restore the P-traps. Next, continue to slip every entice onto its turning into tailpiece. Then continue to tighten your nut, and then swivel it at the manner to satisfy the pipe that goes to the drain. Fix that stop then keep to tighten the nut.

After you are achieved solving the ultra-modern pipes, open the drain and permit the water out to test if there are any leaks. Tighten any connection this is leaking using adjustable pliers.

Tips

If your sink incorporates garbage disposal, the trap is established to it the same manner it's far associated with the tailpiece.

If you can't unscrew the tailpiece, then unscrew the lock nut and push it as a whole lot as eliminate the complete drain. After you replace the drain, unfold the plumber's putty on the clear out's backside detail earlier than changing the lock nut.

Never use pipe dope or plumbing tape to forestall P-trap leaks. None has any impact and will exceptional make the leak worse.

Installing a Kitchen Faucet

Installing a kitchen faucet is easy, and it's a splendid venture for a amateur. It moreover may now not take heaps time. You need first to pick out out out a faucet you decide on. You may additionally furthermore, as an instance, pick out a swish layout crafted from stainless steel.

Faucets in an RV are fixed the identical manner as in houses. You can set up any tap for your RV, even though most RV suitable faucets are plastic, no longer like metallic ones determined in your property.

Tools Needed

- Wrench

- Screwdriver

Method

1. Remove all of the gadgets below your sink so you can get proper of entry to the pipes.

2. Shut off the water supply to keep away from making a large number as you parent. Grab some towels as properly. Some water will live in the pipes even after you turn off the water.

3. Locate the 2 water strains. They can also either be at the water close-off valve or the once more of the cupboard. If yours is installation to the vintage faucet, you want to de-be a part of them to get rid of the old one.

four. Disconnect the water pipes via unscrewing the use of a wrench. Remove the vintage faucet by means of using lifting it after which pull the related traces through the hollow on top of the sink.

5. New faucets normally come with new water pipes already attached, making them easy to put in. The yellow arrow elements to the hose pipe.

6. When you install the tap, the spray hose will make a loop and plug proper lower again right into a clip on the faucet. Ensure the tap is adequately tightened on the spray hose.

7. Place the plastic washing machine this is gray in colour over the housing and the linked lines, after which continue to drop them down inside the beginning on the sink. Remember that your grey is going at the countertop or sink. The relaxation of the hardware stays underneath the sink. Screw the faucet tightly in region.

8. Connect back the water strains under the sink and reattach the spray hose to its clip. Turn at the water and run the faucet, finding out it for leaks. You now have a trendy running faucet.

Chapter 12: Diy Toilet Repairs

Your toilet carries important elements; the upper tank, which holds the water launched whilst you flush the rest room, and the bowl unit, which rests at the ground. The bowl is fabricated from porcelain with out a transferring additives, and truly no longer often does it want safety. The tank, but, is in which most of the renovation happen. You'll be surprised at how easy it's miles to restore relaxation room issues.

Before you embark on any safety, you need to realize how a toilet works.

How Your Toilet Works

The tank above the rest room bowl holds a high-quality amount of water. When you flush your rest room, the water flows down fast through an opening at the lowest of the tank and into the bowl. The pressure pushes the waste from the bowl and into the drain, and finally into the sewer lines.

Two crucial factors inside the tank make this viable: the fill valve or ballcock and the flush valve.

The Toilet Fill Valve (Ballcock)

The lavatory fill valve is the difficulty that fills the tank with water. It is located on the left aspect of the tank and is automatic. When you flush and the water is released, the water stage will fall. It may additionally automatically near off whilst all the water drains into the bowl. The water then rises once more to a selected degree within the tank. The valve is operated via way of a floating cup or floating ball, which movements up and down with the water degree- all of it is based totally upon on your lavatory's layout. The non-floating fill valve operates with the useful useful resource of sensing the pressure of the water at the bottom of the tank.

The rest room in your RV is lighter than regular bathrooms and does no longer have a manage like a domestic relaxation room.

Instead, it has a pedal that is below the bowl. It is flushed through urgent the pedal with a foot, causing sparkling water to flow into the toilet, and a flap opens at the bowl's base. The waste is flushed into the black tank.

Some not unusual rest room issues that a newbie can repair without issues encompass:

- A clogged rest room

- A leak on the rest room base

- A jogging relaxation room

- A faulty flush valve

- A ballcock that doesn't characteristic well

Below, discover three styles of protection that you could do with the aid of using your self.

Clearing Toilet Clog With a Toilet Plunger

An overflowing rest room due to clogging may be disturbing. Such a clog can cause the water to replenish the rest room and spill over onto your floor. This is normally no purpose for panic. It's a reasonably to be had mission.

Before you discover ways to unclog a relaxation room, it's far going to be a wonderful concept to recognize what causes it to clog.

How does the bathroom get clogged?

There's a P-lure under your rest room that holds standing water that acts as a lure and moreover prevents sewer gas odors from filling up the air. Sometimes, objects get trapped within the P-lure, which includes rest room waste or some other foreign places object.

A relaxation room clog can fast be cleared the usage of a toilet plunger. If that does not art work, then use a lavatory auger.

Tools Needed

- Toilet plunger

- Closet plunger

Method

1. Use A Toilet Plunger (Also Called A Closet Plunger Or A Flanged Plunger)

This shape of plunger looks as if the cup plunger but is barely one-of-a-type in form and layout. It has a cup with a tender rubber flap that fills out from within the cup. The flap fits flawlessly over the curved rest room drain, imparting the perfect suction required.

The suction stress in a relaxation room plunger is powerful sufficient to unclog an RV lavatory if the clog is near the top. If the clog is plenty deeper down within the black tank, a plunger will not solve your problem.

This problem may be solved with the resource of pouring boiling water down the toilet bowl and allow it live that way the complete night. You can cut up a cussed clog thru the use of boiling water in a tank that isn't whole but.

Do not use the bathroom plunger to your sink, and do no longer use the cup plunger on your relaxation room. Check out the difference under so you do now not confuse the two:

2. Prepare the Plunger

Add extra water into the rest room bowl just so it's far half of complete or as wished. This is performed to strong a seal throughout the drain setting up.

three. Position the Plunger

Lower the plunger at an attitude into the bowl simply so the cup fills with water. This will boom the plunging pressure. Fit the cup properly over the drain starting just so the flange is inner on the identical time due to the fact the cup seals firmly outside the hole.

3. Pump the Plunger

Use powerful, fast thrusts to push the plunger up and down the drain to loosen the clog. During all movements, hold a great seal. As the plunger comes up, it creates a suction effect to make the clog lose. As it movements down, it pushes the clog down the drain.

Thrust it 5 or six times. When maximum of the water has lengthy beyond out of the bowl, most probably, the clog has cleared.

four. Test it

Flush your rest room to test if it has cleared. If the bowl looks as if it's going to overflow, push down the flapper inside the water tank. This stops the water glide without delay from the tank into the bowl.

Repeat the flushing and plunging gadget until the clog has cleared up.

Clearing The Clog With a Toilet Auger

1. Prepare the Auger

Pull lower back the auger cable in order that the cable's tip is on the stop of the manual tube. Push in the guide tube into the relaxation room, permitting the sweep elbow to rest at the lowest of the bowl and in order that the cable cease reaches farther into the drain organising.

2. Crank the Auger

By cranking the auger's deal with clockwise, push the cable slowly into the rest room lure till it cannot pass any in addition.

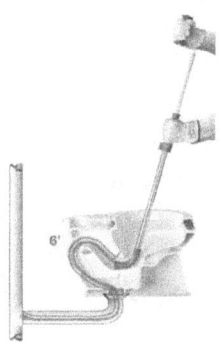

If want be, you could need to contrary the route of the crank to coax the cable thru the curves of your rest room drain. Extend the auger cable till you spoil via the clog.

3. Extract the Cable

As you rotate the address, drag the cable once more slowly out of the rest room. Be moderate as you do this to keep away from scratching the fixture. Make certain the clog has cleared with the resource of flushing the rest room.

How To Fix a Running Toilet

A on foot rest room is one of the not unusual troubles in houses, no longer to say how stressful it is able to be and all the water wasted. Fortunately, it's far very clean to restore this hassle.

Before you start solving your relaxation room, examine a few fundamentals on the manner it truly works.

How the Toilet Works

Understanding the fundamentals of the manner a rest room works will provide you with an idea of what preservation are required. Here are the primary steps:

1.After you press the deal with, a sequence will beautify a flap (flapper), which permits the water in the tank to circulate the bowl. As the tank is emptying, the flapper will drop and near the hole to start the fill up cycle yet again.

2.The go with the go with the flow drops because the tank empties the water. The go together with the float connects to the flow valve, which opens whilst the glide is down and closes even as the tank fills up.

three.At the middle of the tank is an overflow tube, which allows in draining any more water down into the bowl on the identical time because the tank degree receives immoderate. Besides this tube is a channel that the float valve sends water down into the bowl even as it refills all over again.

A lavatory that doesn't prevent taking walks can quit result from severa reasons, which incorporates waterlogged go together with the drift, a immoderate water stage, and a faulty flapper. If none of these is the problem, then most probably, your rest room has a damaged valve. The very last solution is to replace it. Below are techniques to cope with this problem:

1. Check if Your Flapper or Chain Is Faulty

A flapper refers to a cap crafted from plastic or rubber, which permits in maintaining water on your tank. With time, the flapper will become brittle and creates a faulty seal. When your tank cannot hold or refill water, it might be because of a sub-par flapper. The following are procedures to troubleshoot your flapper:

1.Check the consistency of the flapper. In maximum times, the flapper has hardened and stopped growing an unique enough seal. Feel the flapper to set up if it's far no matter the fact that mild and can maintain lower once more the water.

2.Check the chain. If the chain for your water tank is simply too lengthy, trim off the extra thing to keep away from tangling. If the chain is rusty, update it with a present day one.

three.Check if there can be a jam. Sometimes flappers get pinched at the hinges. Adjust this in reality so it works as required.

four.Check for alignment. Flappers can get dislodged, causing leaks. Ensure your flapper seats properly and without delay over the drain.

2. Adjust The Water Level

The overflow tube guarantees that your tank does not overfill and flood the ground. However, if the fill valve is ready too high, there can be a small leakage into the overflow tube and the bowl. The result is a fill valve that mechanically activates, frequently topping up the tank. You can correct this hassle inside the following way:

1.Re-set the go together with the go along with the flow's fill valve. Some valves have steel rods and small clips you squeeze to slide the clip, consequently floating up and down on the rod. If this is the case, circulate the clip right proper all the way down to decrease the water degree. If it's far an antique fill valve with a tank ball and long rod, cautiously bend the rod inside the center to allow the ball to head barely deeper into the tank.

2.Now, flush your rest room and permit it top off and forestall on its very own.

3.The water diploma want to now be approximately ½ to as a minimum one inch below the pinnacle of the overflow tube.

four.If crucial, regulate the float, and flush another time. Do this till the refill stops at the perfect degree.

How To Fix a Flush Handle

This will maximum likely be one of the most effective installations on your rest room. When the flush manipulate is disconnected from the tank or turns into loose, the answers can be one of the following:

- Align the nut in the tank which holds the flush cope with. Tighten it via turning it counterclockwise. Please be aware that the nut keeping your flush deal with from within the tank is threaded in reverse. To loosen the nut way you want to rotate it in the clockwise path. Connect decrease again the improve twine. The very last answer can be to install a brand new flush address.

When fixing a present day-day manipulate, you do now not ought to set up the first-rate type. There are many designs available inside the hardware stores. To understand if a particular deal with will in form your relaxation room, name the customer service and provide them the rest room model and contact and ask them to discover compatibility.

There are modern day handles that healthy nearly any rest room. It does now not keep in mind the shape of lever your rest room has, even though you could require some minor changes.

Another manner to get an appropriate cope with is through the usage of shopping the equal version as the simplest you currently have. You also can call the producer, and they may provide you with options.

Below are the steps to put in a brand new cope with:

1. Open the Tank

Take the lid out lightly and area it apart on a towel, in particular if it is made of brittle porcelain. The control is associated with an extended arm this is connected to a series that lifts the flush valve. Identify the hole that the chain is hooked to and unhook the clasp maintaining the chain to the arm.

2. Remove the Old Handle

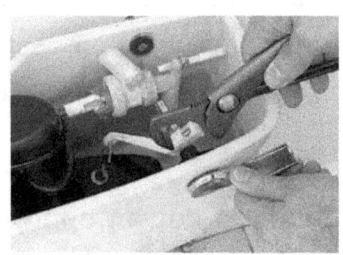

Using a crescent wrench, loosen the nut that holds the deal with in location. Do this with warning because of the reality maximum lavatories have nuts with left-surpassed threads. That approach you want to turn it within the opposite course in choice to turning clockwise like everyday nuts.

Don't use strain whilst loosening a nut because of the reality if you do, it may crack. Add lubricant like WD-40 to the nut if it's miles rusted. After loosening the nut, do away with it through hand and with out issue drift the arm through the hole.

3. Attach the New Handle

With a soapy sponge, scrub off any rust stains or mould that you may find across the cope with hole. Take off the nut from the contemporary deal with, and pop in the arm into the hole. Glide the nut once more over the arm and flip it thru using hand into the lowest of the deal with. Remember, it's far likely a left-handed thread.

Firm it up the use of a crescent wrench, but do not tighten it an excessive amount of, or the porcelain can also crack.

four. Attach the Chain

Clip the chain to the identical hollow that changed into connected to the old arm—take a look at thru flushing numerous times. The flush mechanism want to open and close to sincerely. If the chain is in reality too free, it'll not drain completely. If the chain is without a doubt too tight, it's going to prevent the flush valve from seating flawlessly, causing a non-save you leak.

You can correct this with the aid of using adjusting the chain down or up a hyperlink or . You also can test by means of attempting every other hole the chain can be clipped to. When you are glad the flush is running well, replace the lid onto the tank.

How To Fix a Leaking Toilet

This also can seem like a complex restore to do. However, it's miles quite simple. If you have a look at water round your lavatory base, the subsequent are some of the possible motives:

- The bolts that hold the rest room base to the ground is probably free, causing the rest room to rock, breaking the wax ring seal. If this is the problem, on every occasion the relaxation room flushes, drain water seeps out at the rest room base.

- The wax ring may be faulty. This can be due to the wax ring developing antique and not maintaining the relaxation room firmly at its base.

If possible, do no longer use your leaking relaxation room. The leaking water is usually grimy inflicting nasty odors and capability health risks. Standing water may damage the floor.

Tools and Materials Needed

Depending at the purpose of the water leak, you can require one or extra of the following:

- Bucket

- Sponge

- Tub and tile caulk

- Hacksaw (if wanted)

- Putty knife

- Work gloves

- Adjustable or open-surrender wrench

- Replacement tee bolts

- Replacement wax ring

- Toilet tank insulation

- Toilet tank drip tray

Method

A leaking bathroom can be consistent thru tightening the tee bolts. You can do this via putting off the plastic covers off the tee bolts on each component of the rest room base. With an adjustable wrench, tighten the bolts. This will make the bathroom press down toward the floor, subsequently compressing the wax ring, restoring its function.

If this proves unproductive, the wax ring might be wiped out or damaged. Go to the subsequent step to update the wax ring.

1. Disconnect the Toilet

Start thru way of purchasing the wax ring. Any commonplace wax ring will restore your rest room properly. Shut off the water to your rest room. Do this with the aid of way of manner of final the close-off valve under the left thing of the relaxation room tank. Next, flush the relaxation room. This will drain all of the water out from the tank into the bowl. With a sponge, put off the final water in every the bowl and tank.

Go earlier and unscrew the nut that holds the fill valve tailpiece to the deliver tube. Using an adjustable wrench, put off the tee bolts positioned at the bottom of the relaxation room. If the bolts are corroded and will no longer turn or constantly spin in vicinity, use a hacksaw to sever the bolts.

2. Remove the Toilet

After unbolting, enhance the rest room cautiously and set it on its detail. You ought to probable need to rock the relaxation room gently to interrupt the seal. If you aren't gentle enough, the relaxation room ought to crack. Ask for assist to try this, as toilets have a unusual center of gravity, making it easy to drop them.

When you have got eliminated the rest room, purchase a modern-day set of tee bolts if they have corroded.

3. Remove the Old Wax

Scrap away the vintage wax using a putty knife. Ensure you have got your gloves on, as grim builds over the wax rings as time passes with the aid of manner of.

4. Install the New Wax Ring

Place the new ring at the drain setting up. Ensure the plastic cone faces downwards into the drain. Afterward, healing the tee bolts firmly within the key openings on every factors of the drain flange.

five. Reinstall the Toilet

Cautiously improve the rest room and area it down over the drain, making sure that the tee bolts thread up well thru the holes at the lowest of the rest room. Use your body weight to press down the toilet into the wax ring. Also, rock it gently from one aspect to the alternative to compress the wax, growing a watertight seal.

Put once more the nuts and washers to the tee bolts and tighten using a wrench. Ensure they may be firmly steady but no longer too tight. Remember, if you tighten too much, the porcelain may additionally crack.

6. Reconnect The Water Line

When you reattach the fill valve's tailpiece and the water deliver together, switch on the close to-off valve. If there are any leaks, tighten the connections. When the rest room tank fills up, flush and take a look at at the base to ensure there are no leaks. If the bathroom is not leaking, then your wax ring has correctly sealed.

When you're certain that the wax ring is mounted effectively, run a bead of caulk and bathtub around the relaxation room base on the element the porcelain is in contact with the ground.

DIY Toilet Installations

Toilet installations are not as complex as you might imagine. Some of the installations you may do as a amateur in plumbing embody:

- Installing a fill valve

- Installing a new seat and cowl

Installing a contemporary bathroom isn't always difficult each.

Replacing And Installing a Fill Valve In Your Toilet

When the fill valve to your relaxation room is defective, it is better to update it. Installing a present day fill valve is straightforward enough for any domestic proprietor to do, and a novice can cope with it thoroughly. Follow the stairs under:

1. Shut off the Water Supply

Shut off the water valve that is located certainly under the relaxation room tank. This is the water pipe that protrudes from the water and is established to the rest room tank's underside. Also, close off your crucial water valve determined outside your property. All this is finished to prevent a flood in your property.

2. Drain The Water Tank

Now, flush the bathroom and do not let circulate of the flush lever for some time till all of the water drains. When you have got were given eliminated maximum of the water, use an vintage sponge or towel to absorb the final water at the tank's backside. You can also moreover use a wet-dry vacuum.

three. Detach The Fill Valve

Take off the toilet lid and region it carefully on the ground, preferably on a towel, to save you it from breaking. The valve is the entire plastic column in greater present day lavatories, which incorporates the on-off lever and the float. The flow is a separate element however is joined to the top of the fill valve in older bathrooms. Also, the fill valve has a tube associated with the overflow pipe.

Loosen the nut that connects the water supply to the fill valve to detach it from the rest room. You try this via turning the nut counterclockwise with a wrench or pliers. Pull out the deliver pipe of the valve beneath the tank.

Have a towel handy due to the fact some water might also additionally waft out of the tank whilst you unscrew the nut.

4. Fix The New Toilet Fill Valve

Remove the valve from its packaging, and study all the commands. It comes truely assembled and must be clean to recovery right away within the water tank.

Remember the new fill up tube want to be clipped to the overflow tube.

5. Ensure all The Washers are Properly Fixed

Ensure all the nuts and washers are successfully related at the point wherein the fill valve enters. The washers offer a watertight seal at the connection factors.

Be careful while tightening the nut beneath the fill valve with a wrench or pliers to save you the relaxation room tank or the nuts from cracking, causing a leak.

Chapter 13: Diy Bathroom Repairs

Bathrooms are every other location in your own home in which plumbing issues will upward push up now and again. Some commonplace problems that you can DIY embody:

- Leaking showers

- Leaking relaxation room sinks

- Slow drains

How To Fix a Leaking Shower

We all experience taking our heat showers, especially after a long difficult day at artwork. You additionally love them within the morning to freshen you up in advance than you begin your day. While a leaking bathe may not intervene together together along with your first-rate warm showers, it may increase your water invoice significantly.

It is hard to understand the purpose of a leaking bathe, that is commonly in the decrease again of the wall. You can start through solving the showerhead and speak to a plumber later to find out the actual purpose of the leak. To do this:

Supplies Needed

- Teflon tape

- Old toothbrush

- Distilled white vinegar

- Large bowl

- Crescent wrench

- Pliers

- Rubber O-ring

- Plastic washing device

- Towel

Method

1. Remove the Showerhead

Before you do this, near off the water valve to the relaxation room. You can normally discover it near the bathe or in the basement. You may also moreover decide to reveal off the primary water line. Removing the showerhead may be accomplished resultseasily using your hand.

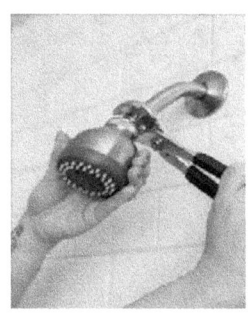

1. Clench the showerhead, and turn it counterclockwise to unscrew it, disconnecting it from the pipe.

2. If it does no longer loosen without issue, hold the bathe pipe with one hand and flip it using your one-of-a-kind hand.

3. If you continue to can't dispose of it using your hands, use a crescent wrench or pliers and cast off it from the wall.

2. Replace The Washers

If the washers or the O-rings are worn out or broken, update them. If they will be in right shape, use Teflon tape to re-wrap the threads at the pipe. The tape seals the gaps the various piping and showerhead.

three. Clean The Showerhead

Do this even as the showerhead is off the wall. Immerse the showerhead in a bowl of vinegar and permit it to sit down down for 2 hours. Afterward, rinse the showerhead in warm water. Take an antique toothbrush and wash many of the nozzles. Use a clean towel to dry the showerhead.

4. Reconnect Back the Showerhead

You can reattach the showerhead using your hand. If this doesn't art work, use a wrench or pliers to tighten until it suits properly. Do no longer tighten an excessive amount of, as you could want to open it frequently to easy it.

Turn on the water and take a look at if the bathe but leaks. If it does, then there is every other solution, that's fixing a leaking faucet.

How To Fix a Leaking Faucet

Fixing a faucet in an RV is sort of similar to that in the home. Since most of the plumbing pipes in RVs are plastic, typically no tools are required as the faucet may be without issue unscrewed the use of your arms. Before embarking on repairing or changing a new faucet, definitely as in a residence, flip off the number one water supply on your RV. This is typically from the freshwater tank.

The way to this problem is to eliminate and replace the cartridge internal the tap.

Supplies Needed

- Adjustable wrench

- New cartridge

- Cartridge puller

- Handle puller

- Pocket knife or awl

- Flat head screwdrivers

- Towel

Method

1.Shut off the water deliver.

2.With a touch stress, use an axe or pocketknife to remove the take care of cap it is within the center of the bathe tap. A screw want to be visible now.

3.Use the screwdriver to remove the cope with screw. If the cope with can not come off without troubles, use a cope with puller.

four. Use a screwdriver to tug out the maintaining clip that rests on pinnacle of the cartridge. The clip looks as if a small horseshoe.

four. Remove the showering machine it in reality is regular throughout the middle of the cartridge. Unfasten the nut and hex screw collectively with your wrench.

6. Pull out the cartridge together along with your pliers. Replace it with a state-of-the-art cartridge whose version is exactly much like the only you've got removed. Follow the commands carefully to align the trendy cartridge efficaciously.

7. Turn to your bathe and check if there are any leaks.

Please look at you've got have been given completed all you could to repair your leaking showerhead. If it maintains to leak, then you may call in a professional plumber.

How To Fix a Slow Sink Drain

This is a very not unusual hassle in the relaxation room. The sluggish drain's number one cause is a clog up by the hair, cleaning soap, and awesome overseas debris.

The sink for your RV can enjoy an overflow hassle just like that in a house. They moreover get clogged due to hair, grease, soap, and so on., stopping the water from draining without issues. For this, a plunger will unclog it, and if it does now not paintings, you may use drain cleaners which can be steady for plastic pipes.

This is a hassle that is straightforward to repair if it has came about in our houses. There are 5 techniques to remedy this hassle, as defined underneath:

1. Remove Debris Using a Zip-It Tool

One manner you can get rid of a slow-draining sink is via way of the usage of a Zip-it tool. This is an an awful lot much less costly device that works wonders in a few minutes. When you slip the Zip-it tool in the drain, it will lure particles, along side hair, and then cast off it out. Repeat till the drain is plain.

2. Remove the Pop-up

If the Zip-it device did now not easy out all of the debris, otherwise you do not have one, then you could exercise this technique. Simply eliminate the pop-up, as a way to have some debris like cleansing cleaning soap stuck onto it. Thoroughly easy it and region it back inside the drain. There's a pop-up nut it's miles inside the back of the sink, that you press right right right down to deliver it out.

three. Use a Homemade Cleaner

You do not need to use harsh chemicals to smooth your drain. You can use homemade remedies which are inexpensive and artwork genuinely as properly. A domestic made cleanser gets rid of the closing debris after the bigger particles were pulled out the usage of a Zip-it tool or cleaned out a sink pop-up.

First, pour boiling water down the drain. Afterward, combination ½ cup of vinegar with ½ cup of baking soda, and allow glide down the drain. Wait for 10 mins and pour boiling water down the drain another time. Finally, flush the drain with everyday warm faucet water.

four. Clear the Sink Overflow

The sink overflow drains water if it by twist of destiny rises too high within the sink. It also lets in air into the drain, helping the water to float quicker into the drain. However, debris can constructing up through the years inside the sink overflow, causing the sink to empty slowly. It is probably notable if you wiped clean the overflow frequently.

The following are steps at the way to easy your sink overflow:

Supplies Needed

- Two or 3 quarts of boiling water

- A sink-cleansing brush or pipe cleaner

- A tube made from silicone or some different non-warmth accomplishing cloth

Method

Push the sink-cleansing pipe into the overflow, interior and out frequently. That will improve out the gunk and construct-up.

After you're extraordinary that all debris is out, stick the funnel into the hole. Pour boiling water via it and into the overflow. This motion will loosen any leftover gunk and debris. Be cautious as you try this to prevent scalding your self with the boiling water. Repeat as required.

Fix the Slow Drain Using a Plunger

Take a cup plunger and cowl it over the sink drain. Plunge it numerous times to loosen and dislodge hair or each other distant places debris. For greater powerful effects, use duct tape to cowl the overflow to create a seal after which plunge. More debris may be cleared out this manner.

DIY Bathroom Installations

You do not have to call a plumber for easy bathroom installations that you can do via your self. Sometimes we call in a plumber to carry out truthful installations, no matter the truth which you have no plumbing revel in.

Some easy installations you can cope with encompass:

- Fixing a tap aerator

- Replacing a unmarried cold and heat bath knob

- Replacing a tub tap deal with

How To Fix a Faucet Aerator

When the water coming out of a tap may be very low, the solution is quite easy. The faucet spout has at its surrender a screw-on show becoming referred to as the aerator. Surprisingly, no longer many human beings realise that there is this form of becoming, and on many activities, they may name in a plumber.

The Function of The Aerator

The aerator works via breaking aside the robust movement of water and provides air to the go together with the waft. This precise function reduces the use of water via 30%. Faucet aerators can get clogged up with mineral construct-up and grit, a not unusual trouble in regions with heavy mineral content material material cloth in the water supply. This problem can be fixed by the use of cleaning the aerator, however on occasion the nice solution is to replace it.

Tools Needed

- Masking tape or rag

- Tongue-and-groove pliers

- Hairdryer (as wanted)

- Penetrating oil as needed

Method

1. Unscrew By Hand

Try unscrewing the aerator from the spout the use of your hand. Most tap aerators are threaded thru hand; consequently, you may unscrew inside the equal manner. Ensure earlier than you start to dry off your palms and faucet to get a corporation grip.

2. Unscrew Using Pliers

If it's far no longer viable to eliminate with the aid of the use of hand, use your pliers, and if you want to reuse the aerator, you can use shielding tape or rag to wrap throughout the aerator to guard the metal floor from scratches. Once you do that, it's miles going to be steady to apply your pliers. Tongue-and-groove pliers are the great for this task.

Clutch the aerator among the plier and jaws, doing it carefully to make sure the jaws live on the aerator, in choice to the tap spout. To unscrew the aerator from the sprout, flip in counterclockwise.

If that also does not paintings, make 1 / 4-turn collectively along with your pliers, and unscrew the aerator from its new characteristic. Continue seeking to loosen it from one in every of a kind positions till you prevail. However, do no longer maintain close to too tightly because of the truth the steel is tender and can bend without troubles, making your undertaking more difficult.

3. Heat and Penetrating Oil

If it's far though difficult to loosen, exercise warmth using a hairdryer. That will make it extend the metal, making it clean to loosen with pliers. If warm temperature fails, you may moreover use penetrating oil at the threads after which try and loosen it with pliers.

When you got loosening it, skip ahead and clean it or installation a present day aerator.

four. Cleaning and Reinstalling The Aerator

After disposing of the aerator, take a look at the steel display. If it's far clogged with mineral deposits or grit, you could easy it off by using manner of hand or use a needle or pin to poke the steel show's openings.

If the aerator is clogged with lime mineral deposits, use a business lime-remover product, or soak in a unmarried day in vinegar.

five. Replace Aerator if Rusted

If it's far rusted, you need to replace it.

When reinstalling or putting in the aerator, screw it returned along facet your hand to first take a look at it. If it leaks, then tighten the aerator a piece extra with pliers. Don't overlook about to use the protecting tape or rage to shield the steel floor.

How to Replace Shower Knob

You may additionally moreover want to alternate or replace your shower knob for aesthetic motives due to the truth it is worn out or wobbling at the same time as establishing it. This is each different easy hobby that a beginner in plumbing can do.

Chapter 14: Diy Plumbing Repairs Outside Your Home

Plumbing safety accomplished outdoor your own home, are without a doubt as important as those finished inner your own home. You additionally want a lovely garden or outside thru watering the grass, plant life, or plants. That is possible if the plumbing for your compound is in proper running situation and nicely maintained.

Draining Your Swimming Pool

This is a charming venture for a newbie. If you have got a pool in your home, that is an easy DIY plumbing mission. Draining your above-ground or in-floor swimming pool isn't tough. However, it is extra related to than actually beginning a drain and letting the water out.

It's critical to empty your pool to maintain the great of the water. You can also want to empty the pool in case you want it repaired. It is also an notable idea to drain your pool if the water has stayed for lengthy with out changing and after a length of heavy rain.

Where want to you drain the pool water?

Sanitary Sewer

This will be the maximum inexperienced manner to drain pool water in case your community permits it. You may want to require a hose pipe linked to the sewer thru the hurricane drain.

Gutter

You can use gutters which sooner or later motive the number one sewer.

Irrigation

If your pool water is right for plants, your municipality can can help you drain water into your own home.

Tools and Materials Needed

- Shop vacuum

- one hundred and fifty-foot nylon rope

- Pool skimmer

- Pump discharge hose

- A Submersible pump (condominium)

Method

1. Rent a 2-inch submersible pump with a capacity of as a minimum 60 gallons normal with minute from your house middle. Also, lease a discharge hose long sufficient to reap the final discharge issue.

2. Screw the submersible pump to the release hose. Spread out the release hose to its complete length, and lay it out flat on the ground to expel the water.

three. Your submersible pump want to appreciably put off particles as lots as ½ inch in diameter. Nonetheless, eliminate as lots junk as you may above that degree.

Afterward, turn off and unplug the pool clear out.

four. Submerge the unplugged pump into the deepest a part of your swimming pool. Do this through lightly letting down the pump with a rope looped thru the manipulate on the pump. Ensure the wire does no longer get pulled into the pump.

five. Plug the pump proper into a ground-fault interrupter outlet. Monitor the pumping operation. Do now not run the pump when it's miles dry as it can get damaged. The pump should be transferring water at all times finally of the way.

6. Turn off and unplug the pump at the same time as the water is virtually too low for the submersible pump, and dispose of it from the pool. With a store vacuum, smooth out the closing water left at the bottom of the pool.

How to Repair Damage on Your Gutter

Having nicely-maintained gutters can prevent flooding in your basement and harm because of moisture. The moisture can weaken the flooring of our domestic and the partitions of the foundation. You can first take a look at for any damages on your gutter, and if there any, you could both restore it or update it.

On the alternative hand, your RV is similar to your own home and additionally calls for gutters to channel away rain, snow, and some other shape of moisture.

Below is a guide on the way to repair the gutters in your home.

Tools Needed

- Plastic spatula/slight tip putty knife

- Scoop device

- Long-dealt with tongs

- Sealant

- Sealant remover

Method

1. Remove Leaves and Debris

Remove the leaves, dust, and particles using a mild putty knife or a plastic spatula. After scrapping your gutter, remove all distinct particles left on the gutter the use of a scoop device. Next, test the elbow joints of your gutters and entries to the downspout for symptoms and signs of clogging. If they've got video display devices, virtually cast off them and clean off the debris and dirt.

If the downspout entries and elbow joints aren't screened by means of way of using the usage of long-handled tongs, take out the debris and leaves.

2. Seal the Cracked Seams

Check the gutters for breaks or leaks, particularly around the seams. This shape of damage may be with out issue repaired with the aid of making use of a water-evidence sealant. First, take away the vintage sealant by way of using a sealant remover and allow it to stand for approximately an hour. Next, scrape off the vintage sealant from the seams, and preserve to permit it dry. Afterward, exercise the cutting-edge sealant.

If your gutters are sagging, then you could pork up them via manner of the use of metal brackets. Do this through way of manner of putting the metal brackets snugly underneath the gutters, using screws to connect them to the steel brackets to your roof.

Check to make sure your downspouts are within the proper function. They ought to be firmly linked to the connectors and securely connected to the partitions of your own home. The water that is going down your downspouts should hit the ground feet some distance from the walls of your home to save you flooding to your basement.

3. Winterize Your Outside Faucet

There's a chance of capability harm to water pipes within the route of wintry weather. To lower the possibility of any harm to the pipes, winterizing is good. This way protective them from freezing and in the long run growing the water and burst the pipes. The harm can translate to lots of bucks at the same time as water soaks into the floors, ceiling, and walls.

A specific element, first of all, is to winterize the taps or hose spigots to your compound. This is an smooth challenge for beginners, in which insulated covers are prepared over the furniture. It is suitable for shielding the out of doors faucets from rupturing when they freeze.

How Outdoor Faucets Covers Work

Faucet covers are available in patterns. Some are crafted from inflexible thermal foam with a flexible gasket at the edges, and others are bendy insulated bags secured across the spigots.

Faucet covers paintings even as warmth is trapped and moves from the indoors pipes to the outdoor spigot. Trapping the warm temperature stops the warm temperature from escaping, preventing the water in the faucet from freezing, growing, and rupturing the spigot.

Materials Needed

- Faucet repair materials (as desired)

- Insulated faucet covers (1 for each out of doors spigot)

Method

Before you begin, make sure you get rid of as an awful lot dust as possible from the spigot.

1. Disconnect The Hoses

Remove any screwed fittings, splitters, and any hoses earlier than the wintry climate starts offevolved. When you do no longer dispose of the hose, it can entice water inner the tap, causing it to freeze. It's usually hard to do that because of the fact human beings will need to apply the water outside until the onset of wintry weather. However, disconnect the hose early to save you the pipes from breaking because of freezing.

2. Inspect the Spigots

Check all your backyard hydrants, spigots, and one of a kind furniture to make sure there are not any drips or leaks. If you discover any issues with them, restore or replace them before the temperatures drop. A leaking washer or cartridge in the spigot will reason it to freeze and later get blocked.

three. Drain the Spigot and Pipes

Shut off the near-off valve to the number one water line that outcomes in the spigot – this need to be the interior one. Let it live open for some hours until all the water in the pipes drains out. Do this once more until you make certain the water drains out truely.

four. Install Outdoor Faucet Covers

The very last step to winterizing outside faucets is with the useful resource of supplying insulation thru the usage of masking them. To do that, honestly set up an outdoor faucet cowl on each spigot and every precise outside fixture.

Faucet covers are particularly dome-formed or rectangular shells prepared over the out of doors taps.

They moreover come in the shape of bendy bags made from thick fabric full of insulation.

Remember to cover no longer handiest your regular spigots however additionally your freeze-proof spigots as properly. That is because even if your freeze-evidence spigots are resistant to freezing, they will be now not without a doubt consistent inside the course of the coldest season.

The tap covers will preserve all of the fixtures heat and dry at some degree in the wintry weather. It is fantastic that taps or pipes essential into your home will both freeze or rupture with this insulation kind. When wintry weather is over, take away the tap covers and keep them for the subsequent wintry weather season. You can use them for many years in advance than changing them.

DIY Plumbing Installations Outside Your Home

Plumbing installations out of doors your private home also can appearance intimidating. However, there are smooth to install projects that you can control yourself and maintain coins. Some of the motives you could require sporting out plumbing installations out of doors your home embody amassing rainwater and controlling flooding even as it rains.

1. How To Install Gutter Brackets

If your gutters are unfastened, you can need to install gutter brackets to stable them firmly. That will prevent your drainage tool from drooping.

Tools Needed

- An assistant

- Drill

- Drill Bits

- Gloves

- Screws

- Ladder

- New Gutter Brackets

Method

Before embarking in this undertaking, you need to make sure your ladder is on a leveled, strong ground. Besides, you need some other person to help you with the useful resource of shielding the ladder and handing you the device. Ideally;

1. Replace your brackets separately.

2. Unhook the bracket from the lip of the gutter.

three. Use a drill for unscrewing the antique bracket from the roof.

four. Your assistant want to be able to come up with a today's gutter bracket and

hook it to the gutter's lip.

five. With your drill, screw your new gutter bracket to the roof via the fascia.

6. If you may be coping with big screws, begin with the aid of drilling at immoderate pace to allow it to pierce the metal. Place some pressure as you drill into the roof.

Tips on Maintaining Your Gutters

1.Clean out your gutters frequently to prevent them from sagging beneath the weight.

2.Use sturdier brackets even as your vintage ones start getting unfastened and susceptible.

How to Install a Rain Barrel

Did you recognize you may preserve in your water bill absolutely thru installing a water barrel? You'd be amazed at how a whole lot water you could harvest from a rain barrel and use it for watering your flowers and vegetation in the garden.

Rain is collected even as water is diverted from the gutter or roof through a downspout and into the barrel. The barrel is exceptional included with a display display screen to clear out any debris. The water can be used not handiest on your garden but is right enough for consuming.

Materials and Tools Needed

- Landscaping material

- Waterproof Sealant

- Hose clamp

- Rubber & metallic washers

- Spigot

- Hacksaw or utility knife

- Drill

- Rain barrel

- Extra diverter or downspout material

- Pencil

- Measuring tape

Method

1. First, determine in which your rain barrel can be located, preferably proper away

beneath a downspout. Place on a raised ground. For this, you may use cement bricks or blocks, ensuring it's miles leveled out flat.

2. Start by means of drilling a hole 2 inches from the lowest of the barrel. This is the hollow from which the water will waft and need to be a piece smaller than the spigot that you can repair.

three. Take your spigot and add the rubber and metal washers to it. Spread a sealant this is water-resistant all through the rubber washing device after which area it in the hollow and clench it firmly for 20 seconds.

4. Add every other steel and rubber washing machine onto the quit of the spigot this is inside the barrel. Some people will add a hose clamp to maintain the spigot in area within the course of heavy storms.

5. Cut a hollow on top of the barrel to feature the get admission to component for the diverter or downspout. Use a hacksaw or software knife to reduce the hole's right

duration in an effort to in shape the diverter properly.

6. Next, drill one go out hollow at the top of the barrel and a few other at the bottom. This lets in greater water to be launched in case the barrel fills up with water. It furthermore releases useless stress.

7. Cut out landscaping cloth an awesome manner to in form over the rain barrel. It is used to prevent leaves, mosquitoes, and one of a kind particles from coming into the barrel. This is completed thru the use of taking off the lid after which putting the reduce material over the barrel. Close the lid. The cloth should stick out of the ends a bit.

8. Cut the downspout. If you may upload a diverter, take measurements, after which noticed off the downspout as required. Attach the diverter and downspout. Secure the connecting tube to the port and stick it in the barrel.

10. Test the rain barrel device with the aid of pouring water into it from the gutter. Use a ladder to get to the gutter, but have someone hold it for you at the bottom. If water does no longer input your barrel, there may be a hollow or blockage inside the downspout or gutter.

How to Replace a Sprinkler Head

Signs that your sprinkler head wishes to get replaced encompass frequent leaks, low water pressure, and terrible spray.

Tools and Materials Needed

- Riser removal tool

- Flexible hose

- Seal tape

- Trowel

- Replacement sprinkler head

- Pliers

Method

1. Shop for the Right Sprinkler Head

You do no longer need an over-watered or under-watered garden. That is why you want to purchase the precise sprinkler head. Your contemporary sprinkler head want to have all of the critical data to find a new matching head.

www.ingramcontent.com/pod-product-compliance
Lightning Source LLC
Chambersburg PA
CBHW071445080526
44587CB00014B/2005